Edited by
Kären Wigen and Caroline Winterer

Time in Maps

From the Age of Discovery to
Our Digital Era

THE UNIVERSITY OF CHICAGO PRESS
Chicago & London

The University of Chicago Press, Chicago 60637
The University of Chicago Press, Ltd., London
© 2020 by The University of Chicago
All rights reserved. No part of this book may be used or reproduced in any manner whatsoever without written permission, except in the case of brief quotations in critical articles and reviews. For more information, contact the University of Chicago Press, 1427 E. 60th St., Chicago, IL 60637.
Published 2020
Printed in Canada

29 28 27 26 25 24 23 22 21 20   1 2 3 4 5

ISBN-13: 978-0-226-71859-0 (cloth)
ISBN-13: 978-0-226-71862-0 (e-book)
DOI: https://doi.org/10.7208/chicago/9780226718620.001.0001

Library of Congress Cataloging-in-Publication Data

Names: Wigen, Kären, 1958- editor. | Winterer, Caroline, 1966- editor. | David Rumsey Map Center, host institution.
Title: Time in maps : from the Age of Discovery to our digital era / edited by Kären Wigen and Caroline Winterer.
Description: Chicago ; London : The University of Chicago Press, 2020. | Papers from a conference held at the David Rumsey Map Center at Stanford University in December 2017. | Includes bibliographical references and index.
Identifiers: LCCN 2019057905 | ISBN 9780226718590 (cloth) | ISBN 9780226718620 (ebook)
Subjects: LCSH: Cartography—History—Congresses. | Time in cartography—Congresses | LCGFT: Conference papers and proceedings.
Classification: LCC GA108.7 .T56 2020 | DDC 912.09—dc23
LC record available at https://lccn.loc.gov/2019057905

♾ This paper meets the requirements of ANSI/NISO Z39.48-1992 (Permanence of Paper).

*For Elizabeth Fischbach
and John Mustain*

CONTENTS

Foreword by Abby Smith Rumsey *ix*
Acknowledgments *xiii*

Introduction: Maps Tell Time *1*
:: *Caroline Winterer and Kären Wigen*

1 Mapping Time in the Twentieth (and Twenty-First) Century *15*
:: *William Rankin*

PART I
Pacific Asia
35

2 Orienting the Past in Early Modern Japan *37*
:: *Kären Wigen*

3 Jesuit Maps in China and Korea: Connecting the Past to the Present *63*
:: *Richard A. Pegg*

PART II
The Atlantic World
77

4 History in Maps from the Aztec Empire *79*
:: *Barbara E. Mundy*

5 Lifting the Veil of Time: Maps, Metaphor, and Antiquarianism in the Seventeenth and Eighteenth Centuries *103*
:: *Veronica Della Dora*

6 A Map of Language *127*
:: *Daniel Rosenberg*

PART III

The United States

147

7 The First American Maps of Deep Time   149
   :: *Caroline Winterer*

8 How Place Became Process: The Origins of Time Mapping in the United States   171
   :: *Susan Schulten*

9 Time, Travel, and Mapping the Landscapes of War   193
   :: *James R. Akerman*

Contributors   219

Index   221

FOREWORD

Abby Smith Rumsey

We all know where we are—nearing the end of the second decade of the twenty-first century. Philosophers may argue that time is a mental construct, but our bodies know better. Time is a place we inhabit. How do we know where we are? The same way mammals and birds do. We process information through the hippocampus, a small seahorse-shaped organ in the brain that maps our perceptions onto a grid composed of cunningly named "place cells." As for time, the body has its own internally generated clock that responds to cues from external stimuli—light, temperature, smell, and so on. From these time and space coordinates, we conjure a mental model of the world by which we navigate our environment.

It is no wonder that all cultures create practices that mark time and space. Humans tell themselves where they come from as a way of knowing who they are and where they belong. Nor is it surprising that their temporospatial imaginations vary greatly across time. At the Time in Space conference, presentations ranged from the late Aztec period to the early modern Japanese and twentieth-century American tourist maps. The fruitfulness of the ensuing conversations, well reflected in this volume, demonstrates how much more there is to know about the past as we learn to read the evidence in map archives.

Historians write about the past from documentary evidence produced by cultures with writing systems whose documents survive and are made accessible through archives and libraries—lamentably, a fractional record of human consciousness. Until recently maps have been marginalized by the profession that trains its experts in textual but not visual analysis. Libraries and archives that serve historians rarely ease access to maps through item-level cataloging, let alone by recording each map in an atlas.

Digital technologies have changed all that. Decades ago, the development of Geographic Information Systems (GIS) heralded a new engagement with mapping and primed our appetite for more. We now have unprecedented online access to cartographic sources through high-resolution images accompanied by rich metadata and an expanding suite of tools for magnification, geo-referencing, overlays, timelines, and animations, among others. Analog maps and globes have been recovered as core historical documents, and historians are learning how to read and interpret them on their own terms, not as derivative or illustrative objects and not just as subjects to be theorized. Like other documentary forms, cartography uses specific tools of representation—compression and scale, color and font, symbology

and iconography, global views and insets—all placed within one frame that reveals what is otherwise obscured by noise. Cartographers' practice of collapsing three dimensions into two—flat maps—or reducing three dimensions to miniature—globes—are brilliant cognitive sleights of hand whereby too much information is rendered legible.

GIS deploys new forms of representation, such as layering data and dynamic mapping. But historians here point out that GIS-derived animated timelines showing change over time emphasize the linearity of a narrative, always within the framework of "before and after." Printed maps are necessarily "static" and allow our attention to linger, dilate, and roam over all the contents within the frame—a very different cognitive mode. These essays argue that such maps are agents of thought, generative of new cognitive modes.

Conference participants were struck by how powerfully maps can represent ideas, things seemingly immaterial that nonetheless leave traces all across the landscape. Ideas wear the guise of metaphors such as veils pulled back to uncover knowledge, trees of time with many branches, rivers of influence and exchange, footprints as synecdoche signifying the traversal of time. They reinforce the notion that our experience of time and space is fundamentally grounded in the physical—our bodies and what they perceive.

The essayists grapple forthrightly with the teleologies of historical maps. In particular they call out our contemporary biases against both sacred chronologies of sin and salvation and secular tales of progress toward enlightenment or Manifest Destiny. The scholars ask what maps can and cannot represent (as opposed to "illustrate"), how they do so, and how they fail. Good mapmakers know that compression and abstraction, if done sloppily, tend to emphasize novelty and technique for their own sake (perhaps colors too bright or type too bold) while failing to accurately represent context or relationships of distance, topography, and contiguity. Examination of failed maps was among the highlights of the conference, incidentally. The presenters offered keen insights into the normally invisible mechanics of cartographic time and space.

Maps establish the context through which we perceive associations. They can prompt us to infer causality, even if the inference is unintended, misleading, or signified something quite different *then* to its audience than it does *now* to us. Data visualization maps, though, such as the justly renowned Charles-Joseph Minard 1869 flow map, *Figurative Map of the successive losses in men of the French Army in the Russian campaign 1812–1813*, are designed to associate effect with cause. By specifying calendar time and local temperatures on the ground, Minard as good as states that the unimaginable loss of life was due to the severity of the Russian winter. The map refrains from suggesting why the campaign was conducted in winter. Its verbal restraint and visual clarity invite us to ponder the horror and draw our own conclusions. Thus it makes what is unimaginable self-evident.

The sheer number of maps under discussion and the fact that they are readily accessible in digital form testify to the extraordinary impact that map archives

have in generating new historical knowledge. This volume of essays frames a series of new questions now possible to address through the marriage of analog and digital mapping technologies.

Map archives at scale create a context for understanding how temporal and spatial tropes travel, spread, disappear or are overwritten, then resurface and are reclaimed for new uses. They provide evidence of how mapping conventions are modified over time through contact and contiguities. Above all, they make clear who decides what gets mapped, who sees maps, who uses them and how.

This volume gives voice to the ineluctable lure of the undiscovered and uncharted. It is thrilling to see these historians walk us up to the edge of the known and lead us into that space where knowledge ends and ignorance begins. Many times during the conference I felt as if we were in a beautiful book-lined library, one that looked old and venerable, but in truth every time someone reached for an atlas on the shelf, a secret door would pop open onto a passageway that led us further and further, deeper and deeper into the archives.

We are once again in an age of discovery. In the nineteenth century we discovered deep time, and in the twentieth deep space. Our sense of who we are and where we come from changed profoundly. As in previous ages of information overload, mapping is once again at the forefront of knowledge representation. In the 1980s David Rumsey began creating a physical and digital map collection with an ever-expanding toolkit of technologies that amplify our knowledge and inspire awe and respect for the genius of maps and their makers. Confident of their long-term value and of the long-term commitment to their stewardship and access, he donated the physical and digital collections to the Stanford University Libraries. Together they built the David Rumsey Map Center both to advance cartographic knowledge and to share the intense pleasure that spending time with maps affords. The similar commitment of libraries with historical map collections—the Library of Congress, the British Library, the John Carter Brown Library, the Osher Map Library, and the Leventhal Map Center, among others—promises an ever-expanding universe of map resources.

Within these essays we spy the beginnings of many paths of exploration and topics for a multitude of similar conferences and books. The paths ahead are yet to be charted. But the territory is open to all of us because the maps are freely available online, well described by rich metadata, and bounded only by the limits of our own curiosity.

ACKNOWLEDGMENTS

The essays in this volume emerged from a conference at the David Rumsey Map Center at Stanford University in November 2017. Our first and most effusive thanks go to David and Abby Rumsey. As of 2016, when the new Rumsey Map Center opened its doors, their extraordinary collection of maps, long accessible online, became available in physical form to researchers from around the world. The beautiful setting of the Rumsey Map Center provided a stimulating gathering place for a conference that ended up attracting over a hundred participants, including both scholars and the general public. We are immensely grateful to the Rumseys for their encouragement and their intellectual contributions to this project both during and after the conference. We thank them also for their generous publication subvention, which allowed the maps to be reproduced here in color.

Two people played an especially important role in getting this project off the ground. Mary Laur, our terrific editor at the University of Chicago Press, was an early and vigorous supporter of the project. She attended both days of the conference and offered wise—and pragmatic—counsel on how to mold the conference papers into important and groundbreaking published essays. Warm thanks also go to Professor Neil Safier, director of the John Carter Brown Library at Brown University, who traveled across the country to attend the conference, and who provided the participants with insightful feedback during the post-conference debriefing session.

Many units at Stanford pooled resources and staff to make the conference and resulting volume happen. Thanks to G. Salim Mohammed, the head and curator of the David Rumsey Map Center, for generously opening the Map Center for the conference, along with Deardra Fuzzell (cartographic technology specialist) and Timothy J. Cruzada (center services supervisor). We are also grateful to Julie Sweetkind-Singer, head of the Branner Earth Sciences Library and Map Collections at Stanford, for her participation in the conference and her intellectual contributions to it. Thanks to Michael A. Keller, Stanford University librarian, for supporting scholarly interchanges such as this one at the Stanford Libraries. The Stanford Humanities Center staff provided valuable assistance in planning and logistics during the conference. Special thanks to Andrea Davies (associate director), Susan Sebbard (assistant director), and Devin Devine (events coordinator). Maria Van Buit-

en in the Department of History managed the conference finances expertly. Two doctoral candidates in the Department of History worked energetically and with seemingly superhuman competence to coordinate a million details for the conference and this volume. To Charlotte Thun-Hohenstein and Charlotte Hull: thank you!

We are grateful, finally, to the Department of History at Stanford University for supporting the research of its faculty in so many ways both large and small. It is a most collegial department, full of intellectually curious, collaborative, and dynamic scholars. We are especially grateful for the department's support for the annual Brilliant Women's History Faculty Lunch, where the editors of this volume first hatched a plan to hold a conference on time in maps. Among the many happy outcomes of those annual lunches, we can now add this volume.

This book is dedicated, ultimately, to all librarians, those generous souls who bring the treasures of the past to life for new generations. We offer this volume especially to Elizabeth Fischbach and John Mustain of the Stanford University Libraries with our thanks and admiration. For decades, they shared the wonders of ancient and modern books and manuscripts with Stanford students and faculty, always with boundless and infectious enthusiasm. The essays and beautiful maps in this book stand as a monument to the scholarship and teaching they encouraged.

FIG. 1.1 The tiny footsteps that begin at right depict the long journey the Aztecs made from their homeland in Aztlan to their imperial city, Tenochtitlan. Unknown creator, "Mapa de Sigüenza, late sixteenth-early seventeenth century." (Detail) Pigment on amatl paper, 54.5 x 77.5 cm. Biblioteca Nacional de Antropología, Mexico, 35-14. Artwork in the public domain, reproduction authorized by Instituto Nacional de Antropología, Mexico.

INTRODUCTION

Caroline Winterer and Kären Wigen

Maps Tell Time

Around 500 years ago, a scribe in Mesoamerica drew tiny black footsteps on green bark paper (fig. I.1). The footsteps wandered around turquoise lagoons and cactus-covered hills, tracing the path the Aztecs took over many years from their homeland in Aztlan to the Valley of Mexico, where they founded the city of Tenochtitlan, seat of their mighty empire.

The Aztecs did not believe that space was a preexisting entity, somehow already there for them to walk through. Instead, they thought that space had to be brought into being through time. All those footsteps, marching over many decades, built the space the Aztecs believed they were divinely ordained to inhabit and rule. In the brilliant colors and shapes scattered across the emerald field, the Aztecs expressed their notion of space and time as unified and mutually constituted. Today the Aztec map lies flat and silent. But in its day, in words spoken as eyes drank in the symbols, the map breathed the world of the Aztecs into life, gave it a physical shape, and positioned it in the historical time of human beings and the cosmic time of the gods.[1]

This book explores how the maps that orient us in space also organize us in time. Drawn from Asia, the Americas, and Europe, the maps considered here point us to the past, the present, and the future, human time and cosmic time. Although many of these maps were created hundreds of years ago, they remind us of what we still feel today: that unlike space, time is maddeningly elusive. We know it is there, but we need material objects to make it real to our senses. In effect, we can only experience time in terms of space: the hands of a clock ticking forward, the pages of a calendar slowly turning, a clarinet propelling sound waves through the air, children growing into adults. Even the most ancient societies developed objects to mark time's passage, from megaliths to sundials.[2] Maps are one of the objects that human beings across cultures have used to give palpable physicality to the passing of time.

This volume is the product of a conference entitled Time in Space: Representing Time in Maps, held at the David Rumsey Map Center at Stanford University in November 2017. We, the editors, are both historians. Kären Wigen is a specialist in early modern Japan, and Caroline Winterer studies pre-twentieth-century North America. In a field still deeply wedded to texts, both of us are highly visual historians. Like a growing number of humanists, we are fascinated by the ways in which maps, diagrams,

drawings, and even buildings and gardens make truth claims that can often seem so real that they disable our critical apparatus by appearing to be self-evident. A serendipitous conversation about our current research projects over lunch one day led to the realization that we were both interested in how some of the maps we were investigating seemed to be as much about time as space. Despite the many miles separating our archives on opposite sides of the Pacific Ocean, we were confronting a similar methodological problem: decoding temporal messages in what we had been trained to treat as a spatial medium. There must be more of us out there, we thought. And we were right.

Capitalizing on the rich collection of maps archived at the Rumsey Center, we assembled a group of leading historians, geographers, art historians, and map curators with specializations in Asia, the Americas, and Europe. We reassured participants that they should not worry about what a map was or was not, since some of us were handling calendars, landscape paintings, and grammatical diagrams that were not maps in the conventional sense of the term. Instead, we urged the group to think of maps as vehicles of cognition, part of a larger world of communicating about space and time in texts, images, and artifacts. We not only tolerated but welcomed conceptual slippage among genres. What we wanted—and what we got—was a conference that showcased the striking imaginative capacities of people from across the globe who turned to the spatial abstraction of maps to confront a variety of temporal questions. The essays here are the fruits of that meeting.

We focused on maps from the past 500 years for the simple reason that it was only after about 1450 that maps rapidly multiplied in quantity, variety, and distribution. Maps have existed since the earliest human societies; some cartographers would say that no human society has ever been truly map-less, for even the fingers of the human hand can become the bays and peninsulas of a makeshift map.[3] Well before the invention of writing, people found themselves drawn to maps, whose spatial abstractions the human mind seems to grasp with uncanny speed. But the period after about 1450 witnessed an unprecedented flowering of maps.[4] The Scientific Revolution, appropriations of classical mapping techniques, the printing press, burgeoning trade routes—all these factors and more drove the proliferation of cartography worldwide.[5] Maps now shifted from extraordinary things—the precious cache of rulers and other elites—to ordinary objects, part of the paper ecology of travelers, soldiers, merchants, explorers, and bureaucrats. The maps in this volume are purposely drawn from a great variety of places in the post-1450 era: China, Japan, Korea, pre-Hispanic Mesoamerica, Europe, and the United States. All show that maps have been flexible instruments for the imaginative exploration of temporal questions that range from the historical (where did we go?) to the existential (where are we going?) and everything in between.

Just as importantly, the conference intervened in a major debate about the value of traditional physical maps in the era of digital mapping. GIS (Geographic Information Systems) has become a powerful new tool for investigating the past because it offers ways to

present large amounts of data in a spatial rather than textual format. The ability to represent historical data spatially has revolutionized historians' research methods. Everyone knows that spatial data lurks everywhere in historical sources: this person was born here, that battle was fought there, and maps have long been the most common illustrative tool historians use to show where their narrative unfolds. But too much geographical information can overwhelm both author and reader. By contrast, the visual medium of GIS lets historians identify and represent complex geographical patterns at a glance.[6]

The new field of spatial history prioritizes geographical space as the narrative frame for charting change. Spatial history is particularly adept at unearthing subtle processes that tend to recede behind spectacular event-driven narratives. Large-scale processes often unfold outside the conscious realization of the people driving or experiencing them. (One philosopher has called the largest of these phenomena "hyperobjects": existing entities like the cosmos or unfolding trends such as climate change that are so massively distributed over time and space that no single person can grasp them in their totality.)[7] To reveal a process—say, deforestation, urbanization, or land dispossession—a historian has to integrate a vast number and range of sources ("big data") that would quickly swamp the narrative capacities of even the most skilled storyteller. By contrast, GIS makes short work of big data.[8] Dramatically plotted in geographical space, either as a static image or an animated sequence that the viewer watches in real time, digital maps allow big processes to move from the background of the historian's narrative to its center.

These powerful tools have emboldened some scholars to argue that traditional maps cannot handle time well. Paper maps are static, they argue, while movement is dynamic; a conventional map shows a snapshot rather than a process. Michael Goodchild, for example, argues that GIS is a better way to capture "flow-like phenomena": complex processes such as army movements that because they combine geographical, temporal, and other information "are difficult to display in map form." A well-designed database, Goodchild argues, is helpful for "extracting, studying, and visualizing different aspects" of historical phenomena.[9]

There are two implicit assumptions in such arguments. The first is about the truth-value of visualization, especially of animated snapshot visualizations. Although they appear to be about space, these visualizations are also fundamentally about time, in two main ways: *historical* (the animation represents a change that happened in the past) and *durational* (we in the present experience the animation in real time, say over one minute). In other words, animations of historical processes use our experience of *durational* time to make an argument about *historical* time.[10] This is new: historical narrative has usually taken the form of prose, and the duration of time for a reader is longer and more diffuse than when we watch a short animation of a process. But images, especially moving images, have great power over us. Since it is the claim of some GIS scholars that these images are better at representing time than paper maps, we should ask in what ways this assertion might be true.

The insights of the philosopher John Dewey are helpful here. In *Art as Experience* (1934), Dewey distinguished between "experience" and "an experience." The first of these, "experience," occurs continuously as we interact with our environments. It is often inchoate, inconclusive, and unmemorable. By contrast, we have "an experience" when the phenomenon we are experiencing feels fulfilling, like solving a mathematical equation, finishing a good book, or being emotionally moved by a work of art. For Dewey, "an experience" is satisfying and self-enclosed, a conclusive and well-demarcated unity set off from the background flow of daily life.[11]

Dewey's observation can be applied to spatial history's animations. Set off from the ordinary run of mere "experience," the snapshot animations of historical data enabled by GIS easily rise to the level of "an experience": they are self-enclosed and narratively satisfying, interpreting with bracing, linear clarity an enormous cache of data. In the space of one minute, we watch kingdoms and forests growing and shrinking over thousands of years.

Still, we must be cautious. First, the processes such animations represent were not necessarily known to historical actors in the way we are now witnessing them: the process may have only been "experience," part of the general background flow of events. It might even have gone totally unnoticed by the historical actors of the time. It is therefore we modern historians who have raised the stakes of our data, ratcheting them in sensory significance from "experience" to "an experience." We can never fully enter into the past, of course, and no historian would argue that we should ignore past processes simply because historical actors were unaware of them. That said, we should approach our modern methodologies and genres with eyes open, aware of their distortions.

The second feature of most snapshot animations is that they move in linear time. You can play the visualization forward or backward. But either way, the framework is one of Cartesian or Newtonian time: the linear, continuous, universal, and endless conception of time developed in Europe during the Scientific Revolution that has now spread around the world, displacing other temporalities or coexisting with them.[12] The ubiquity of Newtonian linear time in spatial history can constrain its visualizations, limiting our imaginations rather than liberating them to explore the many temporalities that have structured the human experience over the millennia. The art historian Keith Moxey and the classicist Denis Feeney have both shown that historical time—even in western Europe—is not universal, but heterochronic, multivalent, and discontinuous.[13] GIS visualizations of historical processes risk imposing a linear ordering of events onto time periods or places in which the idea of Newtonian time did not exist at all or was not yet hegemonic.[14]

The ubiquity of linear time in GIS should come as no surprise. Computers are deeply embedded in a Newtonian worldview; their inner workings replicate the assumptions of those who built them. In the same way that the new railroads of the nineteenth century funneled travelers into rigid iron pathways that became part of what one historian has called "the industrialization of time and space," so GIS ani-

mations constrain our experiences of temporality to the requirements of their medium.[15]

Maps, by contrast, are open to multiple temporalities, as the essays in this volume show. Over the centuries, mapmakers have worked in many media, including bark, cloth, stone, skin, sand, parchment, and paper. The resulting physical archive offers a set of experiences that unfold in time and place in ways that are fundamentally different from those afforded by a digital array. But each medium makes distinctive time-demands on its users. The small paper map made possible by the print revolution of the fifteenth century was a miracle of compression and convenience. Its graphic imagery could be quickly absorbed by a person on the move, making it useful as a directional aid on the spot. Yet even these handy, portable maps require physical handling—a process that engages the body in real (durational) time. Bodily engagement takes more elaborate forms as the presentation of the map becomes more complex. Atlases, for instance, present their own demands. Binding multiple maps in a single book, atlases can still be read only one map at a time. Meanwhile, the largest maps—those painted or mounted on walls and standing screens—pose a different set of physical challenges. While their outlines might be absorbed at a glance, these maps require sustained study at close range if they are to facilitate the intended imaginative journey into distant times and places. The monumentally scaled maps of Renaissance Italy are a case in point. Painted as frescoes that might cover whole rooms, these grand images were designed to be experienced as part of a comprehensive visual program that typically included historical and religious scenes. Strolling down a decorated corridor, or walking slowly around a room, the viewer would gradually integrate a regional map into a grand narrative of political, military, and spiritual action spanning past, present, and future.[16]

It is our contention that each of these modes of engagement facilitates particular kinds of temporal as well as spatial imaginations. Rather than displacing physical maps, digital animations should prompt us to ponder the unique temporal properties of their pre-digital predecessors.

Five Propositions

The purpose of this book is therefore not to promote paper maps over GIS. The happy fact is that we now live in a world where we have both. Technological change is best understood not as progress or decline but as a series of trade-offs. Our approach here is to interpret the advent of digital mapping as an invitation to explore older maps with fresh eyes. To that end, we propose an analytical approach to cartographic temporality for a hybrid era, one in which physical maps and virtual maps exist side by side. With that in mind, this book advances five propositions.

1. *The production of self-consciously historical maps was a hallmark of the global early modern age.* Working out of archives in the post-Columbian Americas and Edo Japan, the editors were initially struck by the coincidence that historical cartography had flourished on opposite sides of the planet at roughly the same

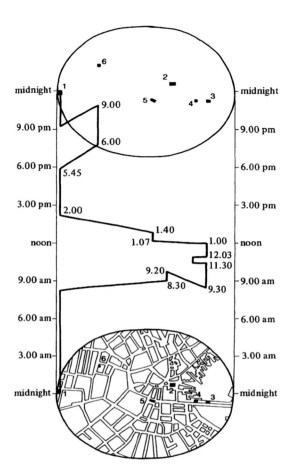

FIG. I.2 This image charts the path of a colonial Boston merchant at various times of the day. From Allan Pred, "Structuration, Biography Formation, and Knowledge: Observations on Port Growth during the Late Mercantile Period," *Environment and Planning D: Society and Space* 2, no. 3 (1984): 268. Copyright © 1984 by SAGE Publications. Reprinted by permission of SAGE Publications, Ltd.

1 home
2 Faneuil Hall
3 countinghouse
4 coffeehouse
5 'change' on State Street
6 club-meeting site

time. In both the Euro-Atlantic and Asia-Pacific region, mapmakers during the sixteenth to eighteenth centuries created distinctive regional repertoires of ways to represent historical events and processes. How to account for that coincidence became a driving question. What was it that compelled people living in the colonized Americas and the Japanese archipelago to turn so often to maps in this didactic way, as a medium for representing the past, during the same centuries?

It is important to note that people in the early modern era did not invent historical cartography. In China, Song dynasty literati were carving ancient toponyms into "earth diagrams" on stone stele at least half a millennium before Columbus.[17] Later, when American and Asian intellectuals found themselves compelled to churn out historical maps, the particular dreams and fears, discoveries and disasters that animated them were not identical. Yet something about the dawn of the global age clearly encouraged people across the world to engage in more and more exuberant historical mapping. The desire to probe that global impulse through comparative sampling—at the colonial margins as well as in the halls of power—shaped this collection from the start.

2. *"Static" maps accommodate time in surprisingly versatile ways.* For hundreds of years, mapmakers have been doing more nuanced and sophisticated things with time than most of us have noticed. The colorful record of the founding of Tenochtitlan that opens this essay (fig. I.1) is just one of many surviving images attesting to an ingenious idiom developed

in the pre-contact Americas for conveying time in a map. The makers of the map created a visual language in which footprints marked the route of major migrations, while dots near a particular toponym (or topo-glyph) symbolized the length of time spent in a given place along the way.[18] A radically different visual language was developed in the 1980s by so-called time-geographers, who added a vertical time-axis to conventional maps in order to show periods of staying-put as well as episodes of movement through a given landscape. In figure I.2, a Boston merchant leaves home in midmorning and spends the next six hours in the commercial heart of Boston before heading home again in midafternoon.[19] Between these two experiments can be found a wide range of techniques and devices for representing spatiotemporal process in a two-dimensional medium.

3. *Diversity persists*. One plausible way to explain these strategies would be to construct a narrative emphasizing global convergence. After all, historical atlases in our own time bear strong family resemblances, no matter where they are produced, since there has been so much sharing of common visual strategies. Today, most atlases display a stock repertoire of familiar icons (dates, arrows, isolines, and standardized symbols for everything from battles to buildings). By contrast, at the start of the roughly 500-year period we are considering here, the cartographic archive encompassed a jumble of practices, a diversity that was enriched in the age of sail by the fitful and uneven dialogue among regional cultures. As late as the eighteenth century, maps created in cosmopolitan settings on different continents still looked strikingly different—such that decoding historical maps from any given pocket of the early modern world demands a significant engagement with local languages and histories. A narrative of convergence from 1450 to the present simply does not tell the whole story when we remain aware of the persistence of diversity. As suggested by the exuberant reproduction, celebration, and display of nonstandard cartography in recent years (whether drawn from the archives or newly created in opposition to the cartographic mainstream), communities worldwide continue to insist on their own distinctive ways of recording time in maps.[20]

4. *All maps tell time*. If the first three claims are essentially historical, this one lies in the domain of epistemology. To be clear, not all maps are purposefully designed to highlight temporal change. Didactic history-maps constitute a relatively small subset of the millions of maps that make up the global cartographic corpus. But to probe how some mapmakers deliberately record the passage of time is ultimately to be confronted with the more fundamental fact that time leaves its mark on all spatial images. Having set out to take an inventory of the purposeful techniques devised by historically minded mapmakers, we have gradually come to see that every mapmaker (and every map reader) has to reckon with time in one way or another. As William Rankin argues in his essay for this volume, all maps carry a temporal stamp—even those that downplay dates or propose a timeless perspective. Some maps revel in the passage of time, while others seem to rise above it, but all these operations (and the sleights of hand on which they

depend) show the centrality of time to what we often approach as a primarily spatial medium. This insight in turn suggests a methodological mandate: in order to truly understand how maps anchor time in space, we need to consider the full cartographic corpus.

5. *Cartographic archives change how maps tell time.* Archived maps can end up telling stories other than those they were originally designed to convey. To some degree, this is a simple function of outlasting their initial purpose. In the same way that an ephemeral snapshot or headline can be transformed into a historical document through the act of filing and cataloguing, so it is with maps. By dint of being collected, curated, and conserved, even the disposable maps that drivers once picked up for free at service stations along the interstate have come to "tell time" in new ways. Like other archived media, such maps come to function less as instruments for living in the present than as clues for reconstructing the past. But the past that they help us reconstruct is inevitably partial. An archive excludes more than it includes; even the most capacious collections, while making room for more than one group's experience, are bound to leave out others. It is when a given archive of maps is canonized—taken as representing an entire community's experience—that those lacunae become pernicious, effectively erasing and overwriting the spatiotemporal imaginations that they leave out.

This kind of erasure happened across the early modern world, wherever imperial conquerors wrote their rivals and victims out of colonial cartography. But something similar played out in an even more pervasive way during the nineteenth and twentieth centuries under the sign of the nation. The ascendancy of the nation-state as the ideal sovereign form prompted official compilation of national atlases, archives, and libraries across the globe. These in turn came to serve as repositories of group memory and sources for self-justifying narratives. While this development is well known to historians, its implications for how we should understand maps—and particularly for our current project of analyzing how maps tell time—have yet to be fully explored. As several of the essays that follow suggest, it was at the moment of national archive creation that maps originally made for other purposes came to be repurposed as founding historical documents for the nation. In fact, only by connecting cartographic dots could many cherished national narratives come into view.

In these ways and more, as the contributions here resoundingly report, conventional cartography remains a vital time-telling resource. The paper map may be physically static, but it is intellectually dynamic in ways that have not been fully appreciated. It is precisely the physically dynamic visualizations of GIS—and the grand claims they have occasionally inspired—that have heightened our attention to the remarkable temporal elasticities of which earlier maps have proven capable. As we hope these essays show, the constraints imposed by dynamic visualizations in GIS enable us to return to early maps with fresh eyes, seeking temporal messages and meanings that we may have overlooked before. In the long run, these alternative temporalities may even help GIS scholars create systems that can better capture the many ways time has been experienced in the past.

The nine essays assembled here offer a starter kit of essential coordinates for charting what is, as yet, largely uncharted intellectual terrain.

Finally, we offer here a road map to the essays in this volume, for readers who want to focus on one geographical region or time period.

The first essay after the introduction is a theoretical salvo by William Rankin that asks how exactly non-animated ("static") maps incorporate time into their design. The author approaches this question as an insider to cartographic practice. Before Rankin became a historian of science and technology, he was a maker of digital maps. Drawing on that experience, he argues that contemporary cartographers have inherited two distinct idioms for representing time, visual codes that reflect fundamentally different notions of temporality. Most maps operate on the principle of a camera, picturing a given landscape at a particular moment in time (whether that moment is defined by a short or long exposure, and whether the resulting image is offered up as a single portrait or arrayed in a time-lapse series). A few maps, however, operate more like our memories, representing the landscape as a cumulative product of notable events: a succession of overlapping layers, each of which has left a mark or deposit that partially obscures its predecessors. Drawing from a wide-ranging archive, Rankin's essay analyzes famous landmarks of American and European cartography as well as more recent maps (including a handful of his own design) to flesh out this taxonomy.

The remainder of the book offers an archival sampler, exploring some of the many modes of map-temporality created during the cartographic explosion of early modern and modern times. The overarching structure here is geographic; the sequence of chapters moves from Japan and China to Europe and the Americas.

Part I parachutes the reader into the cartographic cultures of East Asia, arguably home to the oldest continuous mapping traditions in the world. Kären Wigen leads off with an essay on how historical cartography—which we usually imagine as a development that emerged first in early modern Europe—in fact has a separate history in Asia during roughly the same time period. While some points of connection to Europe fed this new focus—the role of Jesuits such as Matteo Ricci is key—Wigen shows that Asia-specific ambitions, fears, and hopes fed an indigenous practice of historical cartography. Focusing especially on Japan, she shows how the early modern Japanese began to use time-maps to reconstruct nostalgic landscapes of the past. By the 1860s, under the pressure of modernization and Western encroachments, the Japanese turned to cartography to locate themselves in a fast-changing present. Reeling from changes that seemed to be coming fast and uncontrollably, the Japanese turned to maps to orient themselves anew in time and space.

Art historian Richard A. Pegg next broadens the geographical scope with a pan–East Asian perspective on Matteo Ricci's world maps. Originally published at the turn of the seventeenth century (a product of the early encounter between the Jesuit mission and the Ming court), Ricci's imagery had a long if curious career in the Sino-sphere, morphing

over time from an emblem of Western science into a symbol of the first great encounter between the region's great courts and those strange interlopers the Japanese termed "Southern Barbarians." Long after the Jesuits withdrew from the region, Ricci's map lingered. In fact, it would resurface repeatedly for the next two hundred years—seemingly every time a foreign ship appeared in East Asian waters. In Beijing, Seoul, and Edo alike, the return of the Europeans (and their American rivals) in the late eighteenth and nineteenth centuries precipitated particularly marked Ricci revivals. Here, then, is a case of a map whose temporal message fundamentally changed over time. Designed in 1600 to show off the latest in Catholic understandings of the globe and the cosmos, by the 1800s the Ricci map had become the carrier of a subtly different message in East Asia. If not exactly an item of nostalgia, Ricci's map nonetheless echoed through the centuries as a reminder of the culture clash out of which it was born—an echo that acquired alarming overtones during the age of high imperialism.

Following these opening essays on Pacific Asia, part II swings to the other side of the globe to examine three strikingly different cartographies of time, each the product of a distinct cultural context in the early modern Atlantic world. Each chapter in part II stretches the meaning of "map," taking us into one or another outer orbit of the cartographic galaxy. Together they form an important reminder that what we today conventionally call "maps" emerged amid a diverse and creative set of visual experiments. This section opens with art historian Barbara Mundy's close reading of pictographic genealogies from pre-Hispanic Mesoamerica. These are highly stylized images that document the founding of such powerful cities as Tenochtitlan. Decoding the iconography of these Aztec images, Mundy reveals a hybrid genre that combined spatial, temporal, and pictorial elements to relate a bloody history of migration and conquest. With conventions for indicating duration in place as well as movement through the landscape (a landscape whose settlements were likewise represented by conventional glyphs), Aztec mapmakers situated Tenochtitlan's founders in cosmic and human time, construing the capture of the Valley of Mexico as the working out of divine destiny.

Moving across the Atlantic to early modern Europe, Veronica Della Dora's essay explores why the iconography of veils captivated Europeans in the age of global exploration during the seventeenth and eighteenth centuries. Exploration revealed many new wonders to Europeans, but also tested the limits of what they could know about unfamiliar peoples and places. Turning to the elaborate decorative borders and cartouches in maps and atlases printed anywhere from Venice to Amsterdam, Della Dora finds them chockablock with veils, either discreetly draped across or teasingly peeled back from the mapped terrain. Setting these draperies alongside other symbols of time—from the hourglass to the patriarch (Father Time), the winged cherub, and more—Della Dora exposes a rich visual vocabulary that early modern map designers devised for conveying the idea that cartography at once revealed the earth's truths and concealed them.

This section concludes with historian Daniel Rosenberg's analysis of an often overlooked companion of the early modern map: the grammatical diagram. The fruits of an age of exuberant inquiry into languages both spoken and invented, grammatical diagrams emerged along with the Scientific Revolution's impulse to organize and classify. Taking seriously the cartographic significance of prepositions (before, after, under, etc.), Rosenberg approaches grammarians' figures as expressions of time-space relationships at their most abstract. He shows precisely how prepositions came to be arranged in virtual space around an indexical subject. By then setting this visualization in the wider context of early modern European print culture, Rosenberg reveals that geographic maps as we now know them constituted but one genre among many in a crowded field of spatiotemporal imagery. The first great age of printed maps was, after all, also the age of the great encyclopedias, where figures, diagrams, and maps bumped up against each other in a lively visual mélange.

Part III brings us to the United States from the revolutionary era to the twentieth century. Following a loosely chronological sequence, the essays here show that in an almost suffocating cartographic culture devoted to legitimizing the imperial young nation's ever-expanding land claims, other spatiotemporal explorations also emerged. Caroline Winterer opens the section with an analysis of the first maps that attempted to visualize the awesome new nineteenth-century concept of deep time, which substituted a billion-year age for Earth for the traditional 6,000-year chronology set forth in the Bible. Claiming that the "New World" might in fact be older than the "Old World" of Europe, these maps wrestled with the epistemological terrors of time that seemed to stretch into a secular eternity. She also takes up one of the major themes in this volume, which is how these maps of deep time emerged in a highly heterogeneous visual culture that included landscape painting and brain diagrams rather than a hermetically sealed world of images that everyone agreed were "maps."

From deep time we then turn to national time, the subject of Susan Schulten's chapter. Surveying a broad swath of textbooks and historical atlases, Schulten offers a panoramic overview of the chief modes in which successive generations of American historians during the nineteenth century mapped the young republic's history. One of her more striking findings is that innovations in the mapping of census data during the 1850s and 1860s—pioneered in no small part by a wave of skilled German immigrants who entered US government service in those decades—were instrumental in enabling professional historians to develop entirely new theses about the dynamics of American history, especially the existence of a moving "frontier." Without the census's maps, she suggests, it would be impossible to imagine historians such as Frederick Jackson Turner ever generating what turned out to be some of their most celebrated (and controversial) ideas.

The final chapter in part III closes the volume with an extended meditation on the doubled temporalities of battlefield maps. Canvassing the immense archive of American travel books and field guides

aimed at visitors to sites of cataclysmic national conflicts such as the War of 1812, the Civil War, and World War I, James Akerman explores how the designers of these materials responded to the dual imperative of narrating a conflict (telling a story) while leading a visitor through a commemorative site (creating or evoking a personal or national memory). Designed to intertwine storytelling about the past with navigating in the present, the resulting genre was, again, a hybrid—like the Aztec genealogies in incorporating spatial, temporal, and pictorial elements into a single frame.

Together the essays reveal the startling ingenuity with which human beings over 500 years have used maps to locate their collective existence in time and space. In their diversity, they show that beneath the standardizing pressures of globalization in our day, local practices live on. Indeed, deliberate recuperation of regional cartographic idioms offers an important countercurrent to a globalization narrative, if one that is only hinted at here. Map readers across the contemporary globe may be familiar with the same Esperanto, especially as time-telling digital maps now find their way into the palm of nearly everybody's hand. But local communities at every scale, from neighborhood to nation, continue to insist on their own distinctive ways of recording time in maps. We hope that the essays here point the way to new explorations in the cartographic representations of temporality.

Notes

1. For more on Aztec concepts of time and space, in addition to Barbara Mundy's essay in this volume, see Mundy, "Mesoamerican Cartography," in David Woodward and G. Malcolm Lewis, eds., *The History of Cartography*, vol. 2, bk. 3: *Cartography in the Traditional African, American, Arctic, Australian, and Pacific Societies* (Chicago: University of Chicago Press, 1987), 183–256; and Miguel León Portilla, *Aztec Thought and Culture: A Study of the Ancient Nahuatl Mind*, trans. Jack E. David (Norman: University of Oklahoma Press, 1963), 54–57.

2. Kevin Birth, *Objects of Time: How Things Shape Temporality* (New York: Palgrave, 2012); Henri Lefebvre, *The Production of Space*, trans. Donald Nicholson-Smith (1974; repr. Blackwell, 1984), 95ff.; Philip J. Ethington, "Placing the Past: 'Groundwork' for a Spatial Theory of History," *Rethinking History* 11, no. 4 (2007): 465–93; J. L. Heilbrun, *The Sun in the Church: Cathedrals as Solar Observatories* (Cambridge, MA: Harvard University Press, 1999); and Clive L. N. Ruggles, ed., *Handbook of Archaeoastronomy and Ethnoastronomy* (New York: Springer, 2015).

3. See the Micmac "gestural map" in G. Malcom Lewis, "Traditional Cartography in the Americas," in Woodward and Lewis, *The History of Cartography*, vol. 2, bk. 3, 68–69. Denis Wood argues that maps as we know them were only created after around 1500: Wood, *Rethinking the Power of Maps* (New York: Guilford Press, 2010), 1–11. The variety of worldviews expressed in ancient Mediterranean maps is well covered in Richard J. A. Talbert, ed., *Ancient Perspectives: Maps and Their Place in Mesopotamia, Egypt, Greece, and Rome* (Chicago: University of Chicago Press, 2008).

4. Mark Rosen, *The Mapping of Power in Renaissance Italy: Painted Cartographic Cycles in Social and Intellectual Context* (Cambridge: Cambridge University Press, 2014), 3–4.

5. Matthew H. Edney and Mary Sponberg Pedley, eds., *The History of Cartography*, vol. 4: *Cartography in the European Enlightenment* (Chicago: University of Chicago Press, 2019).

6. Anne Kelly Knowles, "GIS and History," in Knowles, ed., *Placing History: How Maps, Spatial Data, and GIS Are Changing Historical Scholarship* (New York: ESRI Press, 2008), 18. For

more recent observations on traditional mapping in the new world of GIS, see Jörn Seemann, Zef Segal, and Bram Vannieuwenhuyze, "Introduction," in Zef Segal and Bram Vannieuwenhuyze, eds., *Motion in Maps—Maps in Motion: Mapping Stories and Movement through Time* (Amsterdam: Amsterdam University Press, forthcoming 2020).

7. Timothy Morton, *Hyperobjects: Philosophy and Ecology after the End of the World* (Minneapolis: University of Minnesota Press, 2013).

8. On these and other advantages of GIS, see Ian Gregory and Alistair Geddes, "Introduction: From Historical GIS to Spatial Humanities: Deepening Scholarship and Broadening Technology," in Gregory and Geddes, eds., *Toward Spatial Humanities: Historical GIS and Spatial History* (Bloomington: Indiana University Press, 2014), ix–xix. For a methodological essay on spatial history, see Richard White, "What Is Spatial History?," Spatial History Lab Working Paper (2010), https://web.stanford.edu/group/spatialhistory/cgi-bin/site/pub.php?id=29.

9. Michael Goodchild, "Combining Space and Time: New Potential for Temporal GIS," and Ian Gregory, "'A Map Is Just a Bad Graph': Why Spatial Statistics Are Important in Historical GIS," both in *Placing History: How Maps, Spatial Data and GIS Are Changing Historical Scholarship*, ed. Anne Kelly Knowles (Redlands, CA: ESRI Press, 2008), 193 (quotation), 123–49.

10. On durational time, see Henri Bergson, *The Creative Mind: An Introduction to Metaphysics*, trans. Mabelle L. Andison (1946; repr. New York: Dover, 2012); and Bergson, *Time and Free Will: An Essay on the Immediate Data of Consciousness* (1901; repr. New York: Routledge, 2013); Anson Rabinbach, *The Human Motor: Energy, Fatigue, and the Origins of Modernity* (New York: Basic Books, 1990), 110–12.

11. John Dewey, "Having an Experience," in *Art as Experience* (New York: Minton, Blach & Co.), 35–37.

12. Donald J. Wilcox, *The Measure of Times Past: Pre-Newtonian Chronologies and the Rhetoric of Relative Time* (Chicago: University of Chicago Press, 1987); and Stephen Kern, *The Culture of Time and Space, 1800–1918*, 2nd ed. (Cambridge, MA: Harvard University Press, 2003).

13. Keith Moxey, *Visual Time: The Image in History* (Durham, NC: Duke University Press, 2013); Denis Feeney, *Caesar's Calendar: Ancient Time and the Beginnings of History* (Berkeley: University of California Press, 2008); see also John Lewis Gaddis, "Time and Space," in *The Landscape of History: How Historians Map the Past* (Oxford: Oxford University Press, 2002), 23–24.

14. Some GIS practitioners are aware of this problem: "The elusive goal is to modify the rigid, Cartesian logic of GIS so that computer-based geographic analysis can more closely mimic the vagaries of human experience." Anne Kelly Knowles, "GIS and History," in Knowles, ed., *Placing History*, 19.

15. Alan Trachtenberg, "Foreword," in Wolfgang Schivelbush, *The Railway Journey: The Industrialization of Time and Space in the 19th Century* (Leamington Spa: Berg, 1977), xiv.

16. Rosen, *The Mapping of Power in Renaissance Italy*, 5; Francesca Fiorana, "Cycles of Painted Maps in the Renaissance," in *The History of Cartography*, vol. 3 (pt. 1): *Cartography in the European Renaissance*, ed. David Woodward (Chicago: University of Chicago Press, 2007), 804–30; and Juergen Schulz, "Maps as Metaphors: Mural Map Cycles of the Italian Renaissance," in *Art and Cartography: Six Historical Essays*, ed. David Woodward (Chicago: University of Chicago Press, 1987), 97–122.

17. See the essays by Cordell Yee in *The History of Cartography*, vol. 2, bk. 2, *Cartography in the Traditional East and Southeast Asian Societies*, ed. J. B. Harley and David Woodward (Chicago: University of Chicago Press, 1994).

18. Barbara Mundy, *The Mapping of New Spain* (Chicago: University of Chicago Press, 1996).

19. Allan Pred, "Daily path of a 'typical' Boston merchant capitalist." In "Structuration, Biography Formation, and Knowledge: Observations on Port Growth during the Late Mercantile Period," *Environment and Planning D: Society and Space* 2, no. 3 (1984): 268. A Wikipedia article on time-geography provides additional examples.

20. On counter-mapping, see the illuminating discussion in Wood, *Rethinking the Power of Maps*, especially chapters 5 and 8.

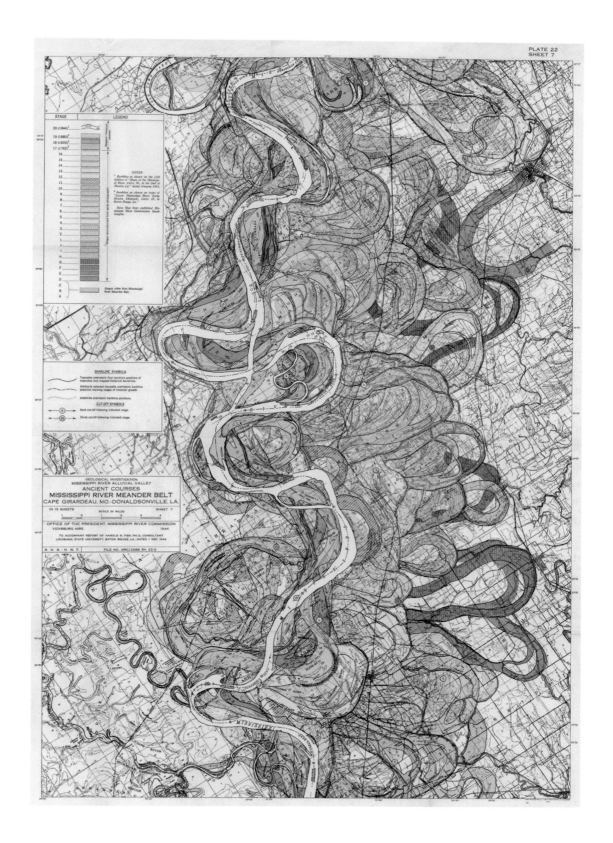

1

William Rankin

Mapping Time in the Twentieth (and Twenty-First) Century

When the Mississippi River overtopped its banks, and then its levees, in the spring and summer of 1927, the destruction was unprecedented. The so-called Great Flood inundated more than 27,000 square miles of prime cropland across ten states—an area larger than the Republic of Ireland—and left more than 700,000 people homeless. It was (and remains) the worst river flood in the history of the United States. It was followed by a decade-long project by the Army Corps of Engineers to radically reshape the river. Over a dozen of the river's famous meanders were cut by artificial channels, and flow capacity was enforced by regular dredging. By the early 1940s, the lower Mississippi was being treated more like an artificial canal than a free-flowing river—it had become substantially shorter, deeper, and faster, and it could handle 50 percent more water than before.[1]

The map shown in figure 1.1 was created in response to this extraordinary intervention. It was published in 1944 after three years of research; it shows the meanderings of the river over the last two thousand years, with each pastel color representing a different historical snapshot of the river. It's stunningly beautiful, with subtle colors and captivating patterns, and it shows the river as a shapeshifting, organic presence—almost alive. This particular sheet shows part of the Mississippi-Arkansas border; the full series of fifteen maps stretches from southern Illinois to southern Louisiana. The lead author was a young geologist at Louisiana State University named Harold Fisk; he had been hired by the Army Corps of Engineers to prepare a comprehensive study on the history of the entire alluvial valley. In his report, Fisk described the Mississippi as a "poised river" that had reached a dynamic equilibrium of constant meandering, and his map offered both a warning and a promise: left to its own devices, the river would undoubtedly revert to its pre-engineered state, but detailed historical knowledge of subsurface conditions and soil types could help the army hold its new ground.[2]

Faced with a map like this, which so elegantly presents a deep temporal story in a clean two-dimensional map—and which has gained something of a cult following of its own[3]—it's rather surprising

FIG. 1.1 The meanders of the Mississippi River over the last two thousand years, with each color showing the river in a different position roughly every hundred years. Harold N. Fisk, *Geological Investigation of the Alluvial Valley of the Lower Mississippi River* (Army Corps of Engineers, Dec. 1, 1944), plate 22, sheet 7.

to come across recent attacks on maps by those of a more digital persuasion. For example, Richard White, one of the early leaders of spatial history, has argued quite plainly that "representations of space cannot be confined to maps" because "maps [are] static while movement is dynamic."[4] One of the champions of Historical GIS in the United Kingdom—Ian Gregory—is even more blunt; he says "a map is just a bad graph" that "cannot handle time," and that research needs to "go beyond mapping" in order to truly understand spatial data. The diagnosis seems relatively straightforward. Gregory says that a map "shows a single slice through time," and therefore "time is fixed." Similarly, the geographer Michael Goodchild describes map layers as but "a single snapshot," while the historian David Bodenhamer agrees that in contemporary mapping, "time is fixed... not dynamic."[5]

Perhaps even more surprising is that these attacks are nothing new. Similar remarks can be found from geographers and mapmakers as early as the 1930s. The German geographer Herbert Knothe, for example, claimed in 1932 that only text could supply "the unmappable factor of time," while John Wright at the American Geographical Society dreamed of replacing atlases of "unadaptable" maps of "static conditions" with some yet-to-be-invented "collection of motion-picture maps."[6] Animated maps have been held up as a panacea since at least the 1950s, and the advent of the computer has led seemingly every discussion of cartographic temporality to draw a hard line between "traditional" paper maps and new forms of dynamic, interactive software that, as Mark Monmonier put it in the early 1990s, "can elevate time to its proper place" and remove the "straitjacket of ossified single-map solutions."[7] As long as there have been alternatives to paper—in film, computers, or the web—static maps have been seen as a dying technology. The vanguard today often disavows mapping altogether, prioritizing user-centered geovisualization and spatial analysis over any outward-facing visual communication.[8]

More is at stake here than just our understanding of how—or whether—maps can represent time. While even the most strident partisans don't claim that maps are *always* atemporal, their complaints nevertheless evoke deep disciplinary prejudices about the relationship between geography and history—or even between space and time in general. Rhetorically, attacking maps means critiquing an outmoded descriptive geography that is only spatial; it also conjures an event-driven history that is only temporal. The embrace of new technology is thus seen as a methodological imperative, a way of repudiating any hard split between spatial pattern and temporal process. Yet in reality, geographers have been interested in temporality for decades, and the intellectual roots of spatial history can be traced not just to the "spatial turn" of GIS—or even the much older "spatial turn" of the neo-Marxists—but deep into the nineteenth century.[9] Ironically enough, denying the temporality of maps ends up conjuring dichotomies (and turf wars) that should have been laid to rest ages ago. Time versus place, past versus present, explanation versus description, linear narrative versus stagnant cartography.

There is a subtle politics here, too. If the pressing issues of our time are about dynamic change—migration, climate, capital flow—and if maps are unworthy of the task, then a methodological imperative can easily become a political one. High-end software and its associated expertise becomes progressive, while maps become conservative, even regressive, despite their broader accessibility.

The goal of this chapter is to show that today's pundits—and their predecessors—are wrong about the temporality of mapping. Harold Fisk's map of the Mississippi is only one striking example among the dozens of ways that static, two-dimensional graphics can tell rich temporal stories. My argument proceeds in three steps. My first task is to analyze how modern maps have in fact shown time, starting from the mid-nineteenth century and remaining mostly unchanged from the early twentieth century to the present. Despite great variety and flexibility, I ultimately find remarkable conceptual coherence: what I call the *photo-cinematic idiom*. I argue that this has been the dominant framework for temporal mapping in Europe and the United States for at least 150 years (I make no claims about other parts of the world). Individual strategies are arrayed along two axes: from the *analytic instant* to *spatial flow*, and from the *single image* to the *collapsed animation*. Seeing maps as nothing but fixed "snapshots" is not correct—but it's not entirely wrong, either, as long as we take the photographic metaphor more seriously.

My second step is to connect this visual analysis of the photo-cinematic idiom to some basic questions in the philosophy of time—especially the relationship between time and space. But rather than suggesting any novel conceptual departure, what I find instead is a rather deep affinity between the temporality of modern maps and the kind of temporality being called for by digital scholars and theorists of geography and history. So it's not just that maps are fully capable of showing time, but that their temporality fuses time and space in precisely the way described by geohistorical theory—and more convincingly, in many cases, than current interactive methods.

Finally, the chapter ends with an alternative. Although most modern maps can be described in photo-cinematic terms, there is one long-standing strategy that is noticeably different—what I call the *historical variable*. While I primarily identify as a historian, I am also active as a cartographer, and this is a strategy that I have explored in my own mapping work and that I have noticed with some frequency in other recent maps as well. Besides offering a useful counterpoint to the mainstream of mapping—a reminder that no cartographic culture is monolithic—historical variables also offer exciting possibilities for exploring what might be described as *spatial memory*.

Characterizing modern mapping as photo-cinematic is more an analytic move than a historical one—it's a way of describing a pervasive cartographic culture, not a claim about how this culture emerged. But the historical implications are nevertheless intentional. Just as photography and cinema emerged in the long nineteenth century alongside a new culture of vision and perception, likewise did most of the specific techniques of photo-cinematic mapping emerge during the same era as part of a

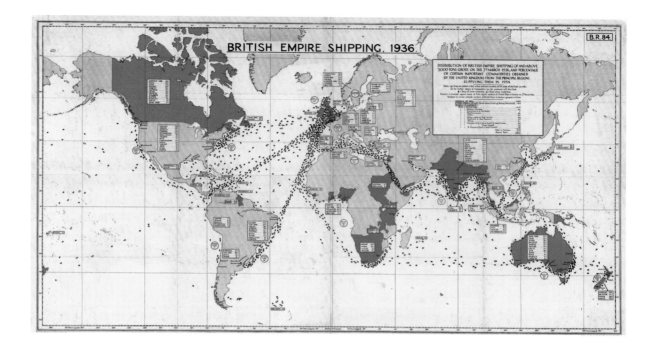

broader rethinking of geographic visibility—one captured by the somewhat problematic term "thematic mapping."[10] In other words, the temporality of photography or cinema might itself be described as *cartographic* just as much as the reverse. A map is rarely a "picture of the ground," and an aerial photograph is not automatically a map, but the kinship is undeniable.[11] This mutuality also means that we should resist any temptation to draw a linear historical arc from, say, the dynamism of film to the dynamism of contemporary computer animation. Photography and cinema are more like fellow travelers than causal agents.[12] Appreciating this historical resonance likewise means rooting a certain approach to time in a specific nineteenth-century culture—one dominated by the concerns of nationalizing states, expanding empires, and their attendant statistics, censuses, and experts. These political concerns provoked many of the new techniques of temporal mapping, and they persist in the background even today.

From Instants to Flows: Defining the Cartographic Moment

The most common critique of maps is that they are merely "snapshots"—the implication being that even the best static map can only capture the world at one particular moment. This may seem straightforward, but in practice there is wide latitude in how the cartographic *moment* can be defined, and this makes all the difference.[13] This is where the photographic metaphor does real work, since the same is true in photography. Actual snapshots must always grapple with shutter speed, and the photographer, like the cartographer, confronts the difference between short and long exposure not as a technicality, but as a strategic choice. Understanding the creativity—and the ubiquity—of the *cartographic shutter* is thus crucial for reading maps' temporality. Moving from the photographic to the cinematic is an important elaboration on this basic approach, but it's conceptually more straightforward.

FIG. 1.2 All the ships in the British merchant fleet on a single day: March 7, 1936. This is a "snapshot"—but the world it shows is hardly static. Published by the British Admiralty (London: HMSO, 1936).

Two extreme examples should make this clear. Figure 1.2 shows a wonderful map of all oceangoing British merchant ships in the world on March 7, 1936. This map was the first of two such maps published in large format by the British Admiralty (the second showed November 24, 1937). The legend explains that "ships are plotted in their actual positions as nearly as the scale of the chart permits," although ships in harbor were only indicated by numbers. In total the map locates almost 2,500 ships around the world, all on a single day. It's perfectly fair to call this a "snapshot," but this is no insult, since it's precisely the narrow temporal specificity of the map that allows it to capture dynamic movement. Notice in particular how the spatial density of ships along certain trade routes acts as a direct measure of temporal flow. The route to Brazil is more trafficked than the route to South Africa, which in turn is more substantial than the route to Australia. Indeed, in 1949 the Harvard geographer Edward Ullman described this map in explicitly temporal terms: it showed "circulation," "movement," and "the pulse of world trade."[14] In other words, the snapshot is not only spatial; instead, it locks space and time together. This is the analytic instant.

At the other extreme, figure 1.3 is a page from the 1894 historical atlas by the influential French geographer Paul Vidal de la Blache. The title is "Ancient Egypt," and like many other maps in this atlas, and similar atlases from the same era (roughly through the interwar period), this single map is meant to show a cultural-geographic situation over a very long stretch of time—in this case more than three thousand years. From our perspective today, it's easy to critique this map for its temporal sleight of hand. The cities and place names on the map were not all current at the same time; their relative importance shifted drastically over the centuries, and even basic regional terms like Lower and Upper Egypt changed their meaning. But these criticisms only highlight how this map in fact makes a strong temporal argument about a *lack* of change over time in ancient Egypt. Indeed, an important part of Vidal's regionalist approach was his interest in precisely those longue durée relationships of human geography later made famous by Fernand Braudel.[15] This is a snapshot of a moment, but the moment is several millennia, and the map again makes claims about space and time together.

In between these two extremes, the most common span for a cartographic moment is one year. This is especially common in statistical maps and atlases that use data collected over the span of a few months—data from censuses, agricultural reports, or social reform surveys. This is how we get maps showing "Leading Religious Denominations: 1950" and "Hogs and Pigs on Farms: 1964."[16] But even here the length of the moment still does analytic work, since these yearlong moments imply that whatever is being mapped is not changing terribly rapidly and that seasonal fluctuations are unimportant.

Maps that show a moment (of any duration) as a frozen image are on one end of the photographic axis; at the other end are maps that are more explicit

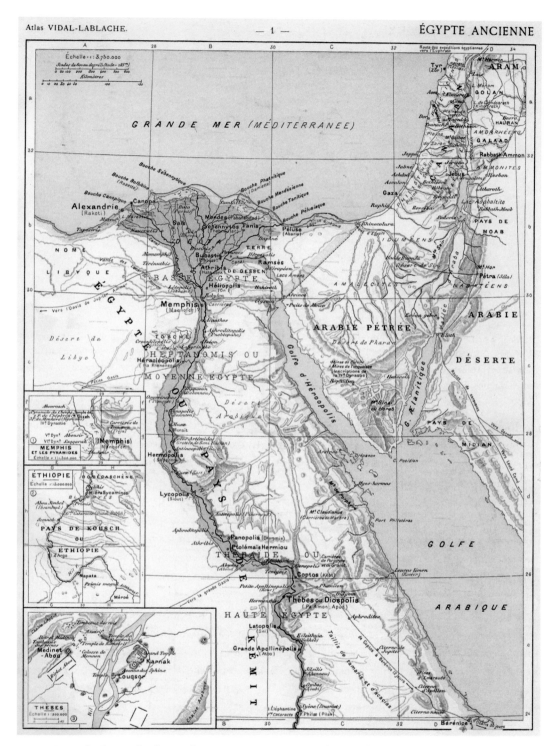

FIG. 1.3 Here, the thousands of years of ancient Egypt are represented as a single, unchanging landscape. This is a remarkably static world, but this temporality is central to the map's argument. From Paul Vidal de La Blache, *Atlas général Vidal-Lablache: Histoire et géographie* (Paris: Armand Colin, 1894), plate 1.

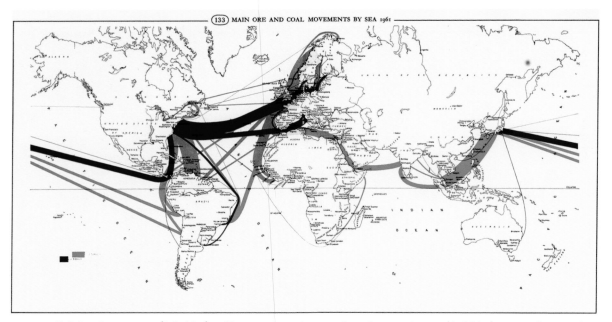

FIG. 1.4 Global trade in iron ore (light blue) and bituminous coal exports from the United States (dark blue) in 1961. This map plots two simple tables of annual statistics, but the flowing lines conjure the actual movement of ships over the span of twelve months. Map by Dorothea Mattern-Authenrieth, from Gustav Adolf Theel, *The World Shipping Scene: Atlas of Shipping, Shipbuilding, Seaports, and Sea-Borne Trade* (Munich: Weststadt-Verlag, 1963), 98–99.

about showing flows, traces, and accumulations. The importance of the durational moment, however, remains the same. Figure 1.4, for example, is a simple example of a flow map from a German shipping atlas published in 1963, where the width of each line symbolizes the tonnage of ore and coal shipped between various ports. Visually, this is essentially a time-smoothed version of the map of British shipping. Instead of representing movement by freezing time on a single day, it represents movement by integrating time over an entire year. This is an important difference, to be sure, but it's not a different understanding of time; in photographic terms, it's simply a difference in exposure time. Just as a long exposure of a highway at night will spatialize movement by producing red and white streaks of various intensities, so too does a flow map symbolize movement through lines of different widths. In photography, there is no categorical difference between short and long exposures—what matters is how the length of the exposure compares to the speed of movement. Similarly, in cartography, frozen maps are no less temporal than flow maps—what matters is how the cartographic moment makes an argument about the rate of spatial change.

Without much effort, it's possible to find maps that show moments of seemingly every duration using a wide range of visual techniques. There are maps that show a day, a month, a year, a decade, a century, and more. Time can be made explicit with shaded areas, patterns of dots, thin traces of movement, thick bands of flow, and arrows of all kinds. Some arbitrary examples: US naval strategy in January 1938; barrels of oil moved per day by Standard Oil; five years of immigration from the British Isles to Canada; thirty-one years of lynchings in the United States; earthquakes and volcanic eruptions throughout all of recorded history; Napoleon's army over a few cold months in 1812 and 1813; the routes of major exploring expeditions from 1838 to 1966. And so on. Not all of

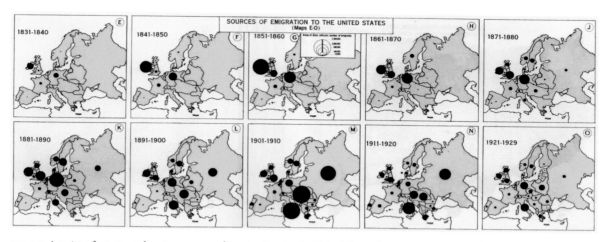

FIG. 1.5 A series of ten maps showing sources of immigration to the United States from 1830 to 1930. In each map, the black circles show the number of immigrants from various countries over a ten-year span; the largest circles represent about two million people. While a graph would more immediately show the three successive waves of immigration (with peaks in the 1850s, 1880s, and 1900s), these maps show regional clustering quite clearly. From Charles O. Paullin (and John K. Wright, ed.), *Atlas of the Historical Geography of the United States* (Washington, DC: Carnegie Institution; New York: American Geographical Society, 1932), plate 70.

these techniques are modern—traces of ships' routes can be found as early as the sixteenth century—but all are relatively common in modern cartographic culture. They appear everywhere, from government reports and classroom wall maps to popular newsmagazines. At times they can be surprisingly complex—recording movement only intermittently, for example—but they require no special training to understand. Academic publications show still further elaborations.[17] But the temporality of all these maps is fundamentally the same: the cartographic shutter is left open for a certain amount of time, and the map records the events, movement, or distribution that occurs during that time. Despite the obvious visual difference between a map that shows a frozen instant and one that shows accumulation or change, the basic understanding of time is unchanged.[18]

Even maps that go out of their way to hide their temporality—maps of an "eternal present" with no explicit temporal markers at all—can almost always be analyzed in similar terms. The generic wall map of the world, the coffee-table atlas, and the topographic quad all make implicit claims about change over time, even if sometimes only through the rhetoric of being "up to date," and even if some of their claims are unconvincing at best. The cues may be subtle and the implications misleading, but the problem is not that maps "cannot handle time." The real issue is that maps can easily make arguments we disagree with.[19]

Beyond this dominant idiom, there are a few allied strategies that lead me to expand the photographic to the photo-cinematic. These are visual techniques that place snapshots in some kind of sequence—but there are several ways this can be done, and the line between photography and cinema is as blurry analytically as it was historically.

The most straightforward is to show two maps—a *before* and an *after*—to highlight change between two points in time. But this strategy need not be limited to two frames. Figure 1.5 is a series of maps from the monumental *Atlas of the Historical Geography of the United States* (1932) showing sources of immigration over a century, from 1830 to 1930. (This is the same atlas that made John Wright imagine motion-picture maps.) Edward Tufte has called these kinds of graphics *small multiples*; instead of showing a simple before

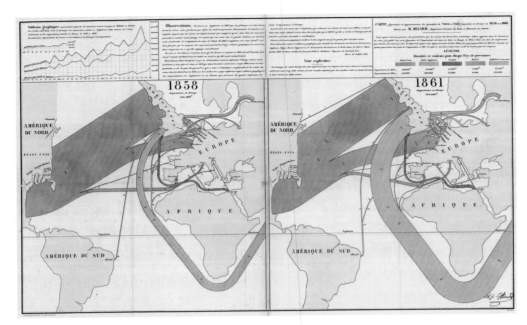

FIG. 1.6 Charles-Joseph Minard's before-and-after flow maps showing how the American Civil War began to affect cotton and wool imports to Europe. Colors show country of origin, with pink for reexports from England. Even though total American exports increased slightly between 1858 and 1861, Great Britain was moving quickly to encourage more reliable sources from its South Asian colonies. Charles-Joseph Minard, *Carte figurative et approximative des quantités de coton en laine importées en Europe en 1858 et en 1861* (Paris, 1862).

and after, the ten maps together show three broad waves of migration—the first in the mid-nineteenth century, mostly from Ireland and Germany, the second in the 1880s, and the third in the early twentieth century from Italy, Austria-Hungary, and Russia.[20] Maps like these tend to underscore the relationship between cinematic time and government statistics, especially the regular rhythm of a national census, and this strategy is often used to support particularly nation-centric narratives—especially those of national expansion, whether territorial or demographic. The obvious photographic analogue here would be something like Eadweard Muybridge's studies of galloping horses from the 1870s, where an overall process is analyzed through a series of static images, and the result is both photographic and cinematic at once.[21]

Just as an individual map can show either a frozen image or flowing movement, so too can a series of maps use any number of techniques. The most famous example of cinematic flow—but certainly not the only one—is probably Charles-Joseph Minard's paired maps of the difference in cotton and wool trade before and after the start of the American Civil War, shown in figure 1.6. Comparing the two maps, each of which accumulates one year of trade, highlights the sharp uptick of imports into England from India (the yellow lines) instead of from the US South—a trend that would only continue as US exports dwindled over the next few years.[22] Minard's graphic shows another cinematic strategy as well: the pairing of a map and a graph. The graph in the upper left shows exports from the United States and India and imports into the UK and France over the three previous decades, starting from the early 1830s. Not only does this provide historical context for the main before-and-after comparison, but it also conjures a continuous series of other maps that aren't shown but could easily be imagined. The overall result, I would argue, is a hybrid illustration of a single spatio-temporal sequence. Map-graph pairs have been used to illustrate all variety of phenomena—economic, demographic, environmental—but in nearly all cases, the relationship is the same: the map illustrates the end of a time series. A spatial pattern is understood as the result of a historical process.

Mapping Time in the Twentieth (and Twenty-First) Century

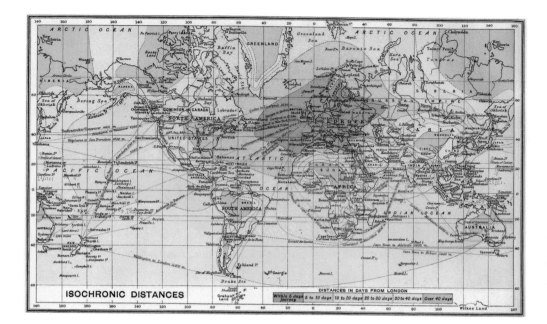

FIG. 1.7 Travel times from London in 1914. Dark red areas are within a five-day journey; dark blue areas are over forty days away. Although these isochronic maps are distinctive for directly mapping time, this same technique of collapsing successive snapshots into one overall composite is quite common when showing spatial expansion, contraction, or diffusion. From J. G. Bartholomew, *Atlas of Economic Geography* (London: Oxford University Press, 1914), plate 12-B.

Finally, the last elaboration is what I call the *collapsed animation*, where successive snapshots are simply overlaid on top of each other as part of a single map. Figure 1.7 shows a classic example: an "isochronic" map of travel time from London in 1914. The sequence of colors, from dark red to cool blue, gives six snapshots, each of which represents the area that could be reached in a given period of time. Although these isochronic maps are often seen as one of few cases where maps can successfully show time, especially if space is distorted into a travel-time cartogram,[23] the same visual technique is used for many other simple trends of expansion and contraction as well—political, demographic, botanical, technological, epidemiological, and so on. Along with before-and-after comparisons and small multiples, collapsed animations are common in statistical or governmental atlases, and they often carry nationalist or imperial overtones. (Isochronic maps, for example, are easy to find before World War I but surprisingly scarce afterward.)[24] Their temporality is stroboscopic, and the best photographic analogy is probably the late nineteenth-century work of Étienne Jules Marey,

where a single negative would be exposed multiple times with a spinning shutter in order to decompose complex movements like jumping, running, or hammering into a composite static image. But Marey is hardly alone. The mid-twentieth-century work of the Albanian-American photographer Gjon Mili, for example—which used a high-speed electric flash and reached from the labs of MIT to the pages of *Life* magazine—likewise shows how powerfully a collapsed animation can dissect complex movements in ways that non-static media simply cannot.

Overall, the basic feature of the photo-cinematic idiom is that a map records a flow of spatial patterns through time. The main question is how to manipulate the shutter: how long to keep it open, whether to take one or a series of shots, or whether to do something more complex like long exposures, multiple exposures, or something else entirely. Conceptually, all photo-cinematic maps—even those that deny their own temporality—could be created by a cartographic camera receiving spatial information over time. And the analogies to Muybridge and Marey are not just circumstantial. Their images are widely under-

stood as part of a modern recalibration of time, motion, and perception. Temporal cartography should be, too.[25]

The Philosophy of the Photographic

The photo-cinematic idiom is the dominant form of temporality in modern Euro-American maps; it is ubiquitous and seemingly straightforward. Yet there is an implicit philosophy of time embedded in these maps, and they suggest a rather specific understanding of the relationship between time and space. This understanding is coherent enough to withstand serious comparison to the more explicit discussions of time by historians, geographers, and other theorists—and the comparison shows remarkable agreement.

Photo-cinematic maps advance two main philosophical propositions. The first is relatively simple: the world consists of a series of successive present-tense moments, and everything has a temporal coordinate that can be treated in similar ways to its spatial coordinates. Every boat on the map of British shipping is located not just on the universal global system of latitude and longitude, but also on the universal timeline of the Gregorian calendar. Similarly, just as objects occupy some amount of space, they also occupy time, and time can be aggregated, abstracted, and generalized much like space. On Vidal's map of Egypt, the spatial location of a city like Memphis is abstracted to a circle, and its temporal location is generalized to several thousand years. Turning time into a coordinate also means that going backward in time and going forward in time are visually indistinguishable. The small multiples of immigration will typically be read from left to right, but they can also be read from right to left with no loss of meaning. (And with some topics, such as plate tectonics, a backward-looking sequence is the norm.) In other words, photo-cinematic time is linear but not inherently directional. It is an abstract mathematical time of simple *succession*, with each frame existing essentially independently of those before and after. It is not the *situated* time of human memory; nor is it the *irreversible* time of causation or entropy.

The second proposition is that time is a *spatial flow*. This means that change over time is registered as change over space, and the world is understood as temporal only because the world changes spatially. This is again directly photographic, since photographic time is likewise inherently spatial: when representing motion, what matters isn't the duration of the exposure itself, but rather the two-dimensional streaks, flows, and blur in the final image.

These propositions can be put into dialogue with two relatively distinct conversations. The first is the digital humanities and spatial history. These fields are the source of the most vocal attacks on static maps, yet nearly without exception what these scholars are actually producing are visualizations of a thoroughly photo-cinematic temporality. For Richard White, and many others, the watchwords of spatial history are *movement* and *dynamic*, since "if space is the question then movement is the answer."[26] This is why interactive applets are seen as a better way to research and represent the dynamic processes of the

past: spatial data literally move on the screen. Yet the visualizations hosted by the Stanford Spatial History Project that White founded—as well as by similar groups elsewhere—come in three main flavors: animations that play a series of snapshots in linear succession; maps that change as the viewer moves a temporal slidebar; and maps that display all events between two dates selected with a double-ended slidebar. These slidebars are especially telling: either they treat time as a coordinate that can be run forward or backward at will, or they act like a cartographic shutter held open for a certain span of time.

I do not want to suggest that these interactive maps aren't innovative, analytic, or pedagogically effective. They clearly are, and I've made similar maps myself. But as an exploration of spatial temporality, they are solidly within the mainstream of modern cartography: time is a reversible linear succession, and spatial change—including movement—is the hallmark of temporality. Philosophically, I see no reason to think that maps that are literally dynamic would make arguments about time that differ in any fundamental way from those of other photo-cinematic techniques. If anything, animations and slidebars seem to be inhibiting experimentation with otherwise common cartographic strategies, including temporal moments other than a year, discontinuous temporalities, or even simple flow lines, traces, and collapsed animations.[27]

The second conversation is the more explicitly theoretical debate about the proper relationship between space and time. The roots here go back to the early twentieth-century philosophy of Henri Bergson and his vehement defense of time as something inherently distinct from space. Bergson proposed his famous theory of duration—*durée*—as an alternative to the kind of mathematical, scientific time—*temps*—that, to him, reduced time to a spatial representation, where time became simply another coordinate, the "fourth dimension of space," and motion became nothing other than a linear trajectory. This was an explicit attack on photo-cinematic temporality, with Bergson singling out Marey's chronophotography in particular as emblematic of the fragmented, immobilized time that he thought could never capture the truth of time—either as actually experienced or as represented in nonmechanistic art.[28] For Bergson, much was at stake in this distinction—causation, free will, even consciousness itself—and his ideas are still remembered today as the main alternative to the mainstream mechanistic-scientific worldview.[29]

For almost sixty years, however, geographers and spatial theorists have lined up against Bergsonism in favor of the kind of integrated space-time totality associated with Bergson's chief public rival, Albert Einstein. The most recent such salvo was made in 2007 by the historian and theorist Philip Ethington in a much-debated article on what he called a "spatial theory of history." Although his ideas continue a long tradition—one that includes Doreen Massey's work from the 1990s and James Blaut's from the 1960s—his article remains state-of-the-art as a philosophical grounding of spatial history.[30] The goal is not just to understand space as historical or as historically constructed, but to understand time itself

as a *spatial* problem, and Ethington's key criticism of Bergsonism is precisely its separation of time from space and its resulting prioritization of time as the key to human life and consciousness. Yet Ethington does not merely offer a defense of space; instead, he concludes that time is in fact *dependent* on space for its very meaning, with time "a mere measurement of spatial motion" that "by itself . . . has no being whatsoever." History, he says, "is not 'change over time' . . . but rather, change through space."[31] Understanding history as spatial thus means taking the spatialization of time—by Marey, or, as Ethington says explicitly, by cartography—quite seriously. Historical time, like cartographic time, is a spatial flow. He even urges historians to see mapping as a guiding methodological metaphor for their craft.[32]

In other words, both in a practical sense with the digital and spatial humanities and also in a more rigorously philosophical sense, scholars today have consistently called for a new spatial temporality, but this temporality turns out to be nearly identical to the long-standing temporality of most actual mapmaking. This is time as a spatial flow, a linear succession of spatial moments where time and space are subject to similar operations of isolation and generalization, and where change in time is only meaningful as change in space. One might even see photo-cinematic mapping as a response *avant-la-lettre* to the methodological debates of the past ten or fifteen years—not only as a rich storehouse of visual tropes, models, and experiments for present-day practitioners, but also as a privileged source for understanding modern temporality quite broadly. It is one thing to say that maps can handle time; this should hardly be controversial. It is quite another to realize that modern cartographic culture shows just how little purchase Bergsonian time has had beyond the academy and how naturalized a four-dimensional space-time ontology has become instead. The jarring difference between historical atlases like Vidal's, where space is but a static stage for narrative history, and atlases from just a few decades later, full of flow lines, arrows, and collapsed animations, only underscores this shift. Maps have *always* channeled shared cultural assumptions about time—even if, for various disciplinary reasons, they have sometimes not been recognized as temporal at all.

The Historical Variable

There is at least one notable exception to the photo-cinematic approach to temporal mapping—what I call the historical variable. It is neither especially new nor terribly complex, but it is noticeably less common and makes rather different claims about space, time, and human experience. Instead of recording whatever events or patterns occur during a certain span of time, a historical variable instead shows the result or summary of a historical process. Visually, the crucial point is that maps of historical variables cannot be created through techniques available in photography, and they are not frames of an animation. As a result, their temporality is not one of simple succession, and rather than channeling four-dimensional space-time, they instead evoke the temporality of human memory.

Mapping Time in the Twentieth (and Twenty-First) Century

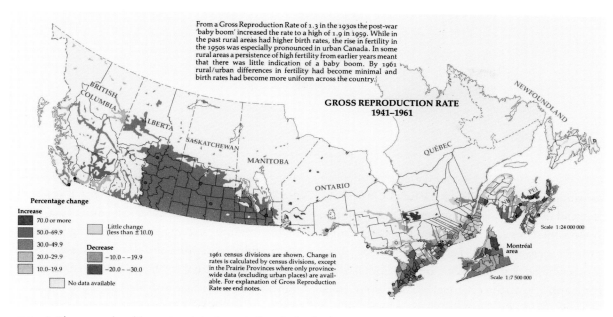

FIG. 1.8 The geography of the postwar baby boom in Canada: big fertility increases in cities and in the west, with steady or even falling rates in the rural east. The variable shown on the map is inherently historical—and inherently directional—since it shows a change over time rather than a particular moment. From Donald Kerr and Deryck W. Holdsworth, eds., *Historical Atlas of Canada*, vol. 3 (Toronto: University of Toronto Press, 1990), plate 59.

The most straightforward type of historical variable records statistical change between two points in time—mostly changes in population, often from government sources. A census atlas, for example, might include a map showing the increase or decrease of the population in various states, counties, or cities since the previous census. Usually warm colors show growth, and cool colors show decline. These kinds of maps first began appearing in the late nineteenth century, but were still rare enough in 1932 that John Wright drew "particular attention" to them when introducing the *Atlas of the Historical Geography of the United States*, which contained only one small series of such maps.[33] Figure 1.8 shows an example from fifty-five years later, when they remained as scarce as ever. The map shows the baby boom in Canada, and the historical variable is the change in the birth rate between 1941 and 1961. The red areas (especially in cities, especially in the West) became much more reproductive during these two decades, while birth rates in the yellow or purple areas (mostly in French Canada) stayed roughly the same or declined. There's nothing terribly complicated going on here; the map simply compares two numbers separated by a certain amount of time and presents the results in diverging colors. But there is no photographic technique that could create this pattern. (And note that this map does not just show change over time, but change in a *rate* over time. The result is doubly dynamic.)

The underlying temporality here is similar to the photo-cinematic, but with one key difference. Time is still a linear spatial flow that can be frozen, captured, and spatialized. But visually these maps are not showing an extended present tense; instead, they show historical change as always already completed. These maps thus show time as inherently directional—usually directionally forward. Compared to a sequence of snapshots that could be run either forward or backward in time, the map of Canada

[28] CHAPTER ONE

has a clear temporal baseline—1941—against which forward-looking change is measured. The order matters, and the result is a narrative, not just a sequence.

In other words, the temporality of a historical variable is always relational rather than absolute—this is why I call it historical. There is always some comparison between different times or some sense of process. Describing the baby boom map, one can only say that the birth rate *increased* or *decreased*; one cannot say what it *is* or *was*. It is this directionality and this clear relationship between past and present that suggests the idea of *spatial memory*. Memories are not simply snapshots of the past, since they must always be recalled in the present and are inevitably colored by the time in between. The archetypal photo-cinematic map is a temporal view from nowhere; it simply captures a certain segment of a timeline. But a historical variable is a situated view; it looks back on the past (or forward to the future) from a specific vantage point. It offers only a partial record and cannot be used to recreate a map of the past.[34]

Seen broadly, this visual strategy in fact has a rather long history, since its other main application is in geological mapping, going back to the eighteenth century. In geological maps, the historical variable isn't a percentile change, but rather the age of the surface. This is how Fisk's maps of the Mississippi work—they show the present situation in 1944, but the surface has been colored to indicate when the land was last part of the river. (Recall that Fisk himself was a geologist.) And notice on his map how older riverbeds have been erased by more recent ones: this is not a collapsed animation, and it could not be turned into a slideshow. Instead, it is a map of environmental memory, placing the past courses of the river into relational time, ordered against each other and viewed from the situated present. With more standard geological maps, it can sometimes be difficult to see any clear historical narrative, since the temporal gaps between neighboring strata are often too vast and geologists' colors rarely show historical sequence. Temporally expressive examples instead tend to show time at the scale of centuries rather than eons—river beds, lava flows—but all such maps are nevertheless part of the same tradition.[35]

In the past few decades, historical variables have remained rare, but they seem to be finding new topics, and recent maps have widened the historical variable beyond the two classic cases of statistical change and geological layering. For example, there are many variables that are historical but do not measure simple change between two points in time. One striking—and happily static—map from the Stanford Spatial History Lab shows the number of times various European territories changed hands during World War II; the result is a map of political memory that highlights the churning chaos of Eastern Europe. One of my own maps uses census data to show the percent of residences in Chicago built before 1939; the result shows the urban memory of the pre-automobile city as seen from the present.[36] Figure 1.9 shows another example from my own work: the trajectories of slavery in the United States. Using data from 1790 through 1860, here the color of each dot represents the decade with the highest

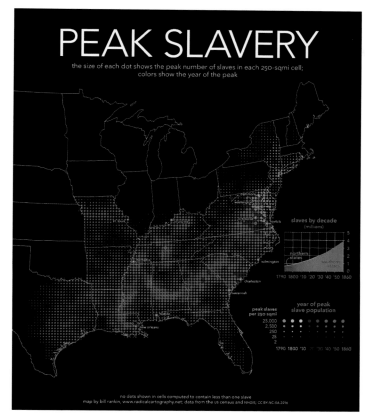

FIG. 1.9 My own map of "Peak Slavery." Here, the historical variable compares the slave population in eight successive censuses, from 1790 to 1860; the colors show the date when the slave population was highest, while the size of each dot shows the number of slaves at that time. Through comparison, the temporality of the map is inherently relational. Posted April 22, 2016, at www.radicalcartography.net/slavery.html.

number of slaves, while the size of the dot shows how many slaves were present at that time. The map thus captures the main historical regions of slavery: the northern states, Delaware, and Maryland, where slavery was already in decline by the end of the eighteenth century; eastern Virginia, Appalachia, and the rice-producing coast of South Carolina, where slavery peaked in the 1820s or 1830s; and then everywhere else, where slavery was still booming at the start of the Civil War. The starting point for this map was a standard series of eight snapshots—one for every decade, each drawn with this same "bubble grid" technique—but when presented as frames of an animation, these historical patterns were all but invisible.[37] One of the common critiques of a "change map" like that of the Canadian baby boom is that any single map can compare at most two points in time and any such comparison hides the original data.[38] This has never been the case for geological maps, and it is not the case for these recent examples, either. The same is undoubtedly true for other historical variables as well.

In most cases, the temporality of memory is only implicit, and sometimes distantly so. But when invoked explicitly, it can transform a map's meaning. The map in figure 1.10, for example, shows the paths of the most destructive hurricanes to make landfall in the United States between 1914 and 2014. For the state of Mississippi, the worst was Katrina in 2005; for Connecticut, it is still the Great Hurricane of 1938. This map shows a comparative variable (similar to peak slavery) in a style reminiscent of Fisk's map of shifting river meanders, with each new record-breaking hurricane erasing the tracks of those that came before. But it's the title that turns a simple number-crunching exercise into a map of spatial memory—one that is pegged to the present day not just with a date range but also with a reminder of the finitude of the human lifespan. Here history is shown leaving imprints in space, and time is not a constant flow of succession but rather a kind of sedimentation. (With geological maps, of course, the sedimentation is literal.) The historical variable does indeed show the past, but the larger goal is to historicize the present.

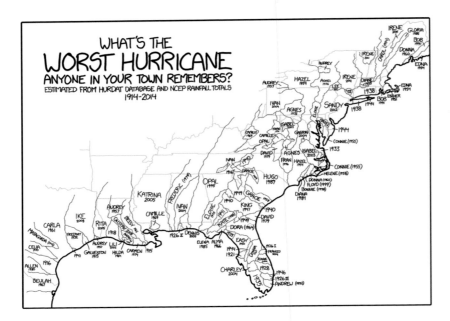

FIG. 1.10 Distilling a century of hurricane wind and rain data results in a map of meteorological memory. Although the historical variable here is comparative, the visual strategy is similar to a geological map. Map by Randall Monroe; posted Aug. 13, 2014, at xkcd.com/1407.

I make no grand claims for the historical variable. It has been used only rarely in the past hundred years, and it remains rare today. But as a real exception to the dominant idiom of modern cartographic temporality, the historical variable makes it clear that maps are not *inherently* photo-cinematic, and capturing a spatial moment by holding open a metaphorical shutter is not the only option. Photo-cinematic mapping normalizes a particular understanding of history: the past becomes a linear succession of spatial relationships, located temporally in a universal timeline and grounded in the kinds of movements and flows captured by governmental, economic, and environmental data. This is history as a cartographic movie—but with more flexibility than usually available in film. The historical variable is a reminder that other understandings of history remain possible, too: history as inheritance, history as narrative, history as groundwork (whether tangible or intangible) for the present. This one exception is enough to suggest that still other versions of history—and still other cartographies—may exist as well, in other eras, in other cultures, or in new visual strategies yet to be imagined.

Notes

1. Brien R. Winkley, *Man-Made Cutoffs on the Lower Mississippi River: Conception, Construction, and River Response* (US Army Corps of Engineers, March 1977; available at dtic.mil), 28–37.

2. Harold N. Fisk, *Geological Investigation of the Alluvial Valley of the Lower Mississippi River* (US Army Corps of Engineers, 1 Dec. 1944; available at lmvmapping.erdc.usace.army.mil), 50. For applications, see Rufus J. LeBlanc Sr., "Harold Norman Fisk as a Consultant to the Mississippi River Commission, 1948–1964—An Eye-Witness Account," *Engineering Geology* 45 (1996): 15–36.

3. See Christopher Morris, "Reckoning with 'The Crookedest River in the World': The Maps of Harold Norman Fisk," *Southern Quarterly* 52 (Spring 2015): 30–44, though I disagree with his interpretation of Fisk's argument. I host the maps at www.radicalcartography.net/fisk.html.

4. Richard White, "What Is Spatial History?," published at web.stanford.edu/group/spatialhistory/cgi-bin/site/pub.php?id=29, quote on 3.

5. Ian N. Gregory, "'A Map Is Just a Bad Graph': Why Spatial Statistics Are Important in Historical GIS," 125; Michael F. Goodchild, "Combining Space and Time: New Potential for Temporal GIS," 194; David J. Bodenhamer, "History and GIS: Implications for the Discipline," 228; all in Anne Kelly Knowles, ed., *Placing History: How Maps, Spatial Data, and GIS Are Changing Historical Scholarship* (Redlands, CA: ESRI Press, 2008). These sentiments are not confined to Historical GIS;

Mapping Time in the Twentieth (and Twenty-First) Century

see Donna J. Peuquet, *Representations of Space and Time* (New York: Guilford Press, 2002), 155; Mark Harrower and Sara Fabrikant, "The Role of Map Animation for Geographic Visualization," 50, 52, in Martin Dodge, Mary McDerby, and Martin Turner, eds., *Geographic Visualization: Concept, Tools and Applications* (Chichester: John Wiley and Sons, 2008); Edward L. Ayers, "Mapping Time," 215–20, and Ian Johnson "Spatiality and the Social Web," 272, both in Michael Dear, Jim Ketchum, Sarah Luria, and Douglas Richardson, eds., *GeoHumanities: Art, History, Text at the Edge of Place* (New York: Routledge, 2011).

6. Herbert Knothe, "Zur Frage der Kartographie als selbständiger Wissenschaft," *Geographische Zeitschrift* 38 (1932): 287; and J. K. Wright, introduction to Charles O. Paullin, *Atlas of the Historical Geography of the United States* (Washington, DC: Carnegie Institution; New York: American Geographical Society, 1932), xiv.

7. See Norman J. W. Thrower, "Animated Cartography," *Professional Geographer* 11 (1959): 9–12; and Thrower, "Animated Cartography in the United States," *International Yearbook of Cartography* (1961): 20–28. Quotes from Mark Monmonier, "Strategies for the Visualization of Geographic Time-Series Data," *Cartographica* 27 (Oct. 1990): 40, and his introduction to Michael Peterson, *Interactive and Animated Cartography* (Englewood Cliffs, NJ: Prentice-Hall, 1995), ix. For precedent to Gregory, note David Unwin, *Introductory Spatial Analysis* (London: Methuen, 1981), 6: "[Maps] are essentially *static* and cannot be drawn to incorporate a time dimension."

8. Geovisualization first emerged in the 1990s; I refer here to recent rhetoric separating visualization from mapping: Michael Peterson, "Animated Map," and M. J. Kraak, "Visualization and Maps," both in Mark Monmonier, ed., *Cartography in the Twentieth Century* (Chicago: University of Chicago Press, 2016); and Todd Presner and David Shepard, "Mapping the Geospatial Turn," 207, in Susan Schreibman, Ray Siemens, and John Unsworth, eds., *A New Companion to Digital Humanities* (Chichester: John Wiley, 2016). Temporal GIS, while important, had relatively little to say about visual representation; see Gail Langran, *Time in Geographic Information Systems* (London: Taylor & Francis, 1992), 25: "Display, while crucial . . . will be left to a more dedicated study."

9. For a useful overview of geographers' temporalities since the 1960s, see Irina Ren Vasiliev, "Mapping Time," *Cartographica* 34 (Summer 1997): 1–50. For a broader treatment, see Alan R. H. Baker, *Geography and History: Bridging the Divide* (Cambridge: Cambridge University Press, 2003).

10. Jonathan Crary, *Techniques of the Observer: On Vision and Modernity in the Nineteenth Century* (Cambridge, MA: MIT Press, 1990); and Susan Schulten, *Mapping the Nation: History and Cartography in Nineteenth-Century America* (Chicago: University of Chicago Press, 2012). But note that "thematic map" is an actor's category of the mid-twentieth century; see Denis Wood and John Krygier, "Map Types," in Rob Kitchin and Nigel Thrift, eds., *International Encyclopedia of Human Geography* (Amsterdam: Elsevier, 2009), 339–43.

11. William Rankin, *After the Map: Cartography, Navigation, and the Transformation of Territory in the Twentieth Century* (Chicago: University of Chicago Press, 2016), 50–52.

12. For this argument, see Sébastien Caquard, "Foreshadowing Contemporary Digital Cartography: A Historical Review of Cinematic Maps in Films," *Cartographic Journal* 46 (Feb. 2009): 46–55.

13. Unlike Vasiliev, "Mapping Time," 11, I do not equate *moment* with an instantaneous "singularity."

14. Cited in Matthew K. Chew, "A Picture Worth Forty-One Words: Charles Elton, Introduced Species and the 1936 Admiralty Map of British Empire Shipping," *Journal of Transport History* 35 (Dec. 2014): 225–35.

15. Paul Vidal de la Blache, "Des caractères distinctifs de la géographie," *Annales de géographie* 22 (1913): 299.

16. Both from *The National Atlas of the United States of America* (Washington, DC: US Geological Survey, 1970).

17. The *Sunday News* of Sept. 2, 1945, includes a trace map by Edwin Sundberg showing every hurricane from the month of September—"Hurricane Month"—between 1877 and 1944. The most well-known academic techniques are the oblique space-time traces associated with Torsten Hägerstrand and his students from the 1970s. The technique did persist; see Mei-Po

Kwan, "Feminist Visualization: Re-Envisioning GIS as a Method in Feminist Geographic Research," *Annals of the Association of American Geographers* 92 (2002): 654.

18. Note that I do not equate *time* with *change*, unlike much of the literature cited above. While it's true that many durational maps do not show change, this does not make them antitemporal. And maps of change are easy to find, using everything from simple arrows to complex layerings.

19. Philip Muehrcke with Juliana O. Muehrcke, *Map Use: Reading, Analysis, and Interpretation* (Madison, WI: JP Publications, 1978), chapter 5; and Denis Wood with Jon Fels, *The Power of Maps* (New York: Guilford, 1992), 126–30. The specific phrase "eternal present" has been applied to maps since the 1990s, but I have not found an originating source.

20. Edward Tufte, *The Visual Display of Quantitative Information* (Cheshire, CT: Graphics Press, 1983), 170–75.

21. Muybridge published still photographs but also presented animations using his "zoopraxiscope."

22. In 1865 Minard published maps comparing the 1858 cotton trade with 1864, by which time the vast majority of UK imports were from India.

23. Vasiliev, "Mapping Time," 13–14; and Mark Harrower, "Time, Time Geography, Temporal Change, and Cartography," in Monmonier, *Cartography in the Twentieth Century*, 1529.

24. The first seems to be Francis Galton, "On the Construction of Isochronic Passage-Charts," *Proceedings of the Royal Geographical Society* 3 (Nov. 1881): 657–58. A notable post-WWI example is in Paullin and Wright's 1932 atlas. After their popular retreat, they remained of interest to academics.

25. Anson Rabinbach, *The Human Motor: Energy, Fatigue, and the Origins of Modernity* (New York: Basic Books, 1990); and Joel Snyder, "Visualization and Visibility," in Caroline Jones and Peter Galison, eds., *Picturing Science, Producing Art* (New York: Routledge, 1998), 379–400.

26. White, "What Is Spatial History?," 3.

27. There are of course exceptions, and I cannot pretend to be familiar with all available work. My assessment is based on projects available online from labs at Stanford, Richmond, Penn, Rice, Yale, and Brown.

28. Rabinbach, *The Human Motor*, 110–12.

29. Gilles Deleuze, *Le Bergsonisme* (Paris: Presses universitaires de France, 1966); and Jimena Canales, *The Physicist and the Philosopher: Einstein, Bergson, and the Debate That Changed Our Understanding of Time* (Princeton, NJ: Princeton University Press, 2015).

30. Philip J. Ethington, "Placing the Past: 'Groundwork' for a Spatial Theory of History," *Rethinking History* 11 (Dec. 2007): 465–93. Both Massey and Blaut reference Einstein, but geographers generally ignore relativistic space-time interactions: Doreen Massey, "Politics and Space/Time," in her *Space, Place, and Gender* (Minneapolis: University of Minnesota Press, 1994), 261; and James M. Blaut, "Space and Process," *Professional Geographer* 8 (July 1961): 2. For Ethington's influence, see Trevor M. Harris, "Geohumanities: Engaging Space and Place in the Humanities," in S. Aitken and G. Valentine, eds., *Approaches to Human Geography: Philosophies, Theories, People and Practice* (London: Sage, 2015), 181–92; and Presner and Shepard, "Mapping the Geospatial Turn."

31. Ethington, "Placing the Past," 466.

32. Ethington refers to John L. Gaddis, *The Landscape of History: How Historians Map the Past* (New York: Oxford University Press, 2002). But I am struck by the similarity to William Norton's spare definition of geographic process: "a set of rules that transforms map forms through time." See his *Historical Analysis in Geography* (New York: Longman, 1984), 26.

33. Wright, introduction to Paullin, *Atlas of the Historical Geography of the United States*, xiv. The first such map that I know of is from Henry Gannett's *Statistical Atlas of the United States Based on Results of the Eleventh Census* (Washington, DC: USGPO, 1898), plate 8.

34. This inherent partiality is not necessarily about the *distortion* of the past, but rather a *loss of information*. Maps of past moments can be combined to create a historical variable, but the reverse transformation is impossible. Human memory, of course, is both distorting and lossy.

35. For lava, see John Auldjo, *Sketches of Vesuvius, With Short Accounts of its Principle Eruptions, from the Commencement of the Christian Era to the Present Time* (London: Longman,

Rees, Orme, Brown, Green, Longman, 1833), facing 26. Auldjo was likewise a geologist.

36. Michael De Groot (map with Erik Steiner), "Building the New Order: 1938–1945," posted Aug. 24, 2010, at web.stanford.edu/group/spatialhistory/cgi-bin/site/pub.php?id=51. Bill Rankin, "Chicagoland," posted Aug. 29, 2006, at www.radicalcartography.net/chicagoland.html.

37. Choropleth mapping by date of peak population is mentioned by Monmonier, "Strategies for the Visualization,"

35. For a good example showing date of initial settlement, see R. Louis Gentilcore, ed., *Historical Atlas of Canada,* vol. 2: *The Land Transformed, 1800–1891* (Toronto: University of Toronto Press, 1993), plate 42. "Bubble grid" is my term; compare to the un-interpolated suggestion by Jacques Bertin, *The Semiology of Graphics* (1967; repr. Redlands, CA: ESRI Press, 2011), 127.

38. Gregory, "'A Map Is Just a Bad Graph,'" 125.

PART I

Pacific Asia

East Asia is an apt place to begin a survey of historical mapping in the early modern world, for nowhere else were history and cartography held in greater esteem. By the sixteenth century, the literate classes in China, Korea, and Japan alike had been steeped in history texts and geographical imagery for centuries. Scholar-officials were expected to be cartographically literate; map-reading and map-drawing featured frequently in their official duties. Moreover, maps were part of a fluid array of techniques for representing place, and mapmakers drew freely on a range of visual idioms, from landscape painting to abstract diagrams. The arrival of map-bearing Europeans in the sixteenth century did not literally reset East Asian calendars, but it did open whole new worlds of understanding, and forced East Asian literati to reorient themselves in time and space. The result was an explosion of historical cartography across Pacific Asia.

Kären Wigen opens this section by canvassing developments in Japan, where three distinct traditions came of age during the Pax Tokugawa (1600–1868). One group of mapmakers emphasized imperial time. By noting earlier political boundaries, former capitals, and long-lost toponyms, cartographers working in this state-centric tradition put the antiquity of the realm on display. Meanwhile, a second group of mapmakers set their sights on more local horizons. Eager to excavate an illustrious past for their own native places, regional history buffs engaged in a burst of amateur cartography, mapping ancestral sites in hand-drawn manuscripts that circulated among friends. Finally, a third way of marking temporal change on maps arose in the flourishing print marketplace, where publishers sought to satisfy the public's hunger for news without alarming the censors. Playing up place names associated with disasters, insurrection, or popular uprisings—all taboo topics for publication—canny cartographers in the turbulent nineteenth century started putting out bird's-eye views depicting the setting of events that could not otherwise be discussed.

Chinese and Koreans also turned to maps to grapple with challenging times, albeit in their own ways. Richard Pegg homes in on a striking set of compound images that appeared in China during the final decades of the Qing dynasty (1644–1911): scrolls and prints that featured a contemporary globe alongside cruder world maps drawn in a style popularized by the Jesuits two centuries earlier. Why, he asks, would a sophisticated nineteenth-century mapmaker fill the

margins around a state-of-the-art globe with quaint but dated imagery? For a clue, he turns to the titles of these works, which pointedly yoked the present ("here/now") with an unspecified past ("there/then"). Evidently, for these mapmakers, the relevant past was the Jesuit moment—and maps created during that extended seventeenth-century encounter with the West proved "good to think with" about contemporary geopolitics.

2

Kären Wigen

Orienting the Past in Early Modern Japan

We are always mapping the invisible or the unattainable or the erasable, the future or the past, the whatever-is-not-here-present-to-our-senses-now and, through the gift of maps, transmuting it into everything it is not...
—DENIS WOOD, *Rethinking the Power of Maps* (2010)

Introduction

Historical mapping today is a globe-spanning phenomenon. Across the twenty-first-century world, sophisticated cartographic illustrations accompany most history textbooks, and they tend to bear a broad family resemblance. Typically, a simplified topographic base is topped with a layer of past place names—chosen for their relevance to a given theme and time—which in turn may be overlaid with graphic elements from the repertoire of contemporary cartography (flow lines, arrows, icons) to capture the geography of a particular problem or process from an earlier moment or era. A specific slice of time is targeted, whether thick or thin, and its span on the timeline is pegged in the map's title (or legend). In these ways, conventional historical cartography today functions as a kind of visual Esperanto. The development of this global visual style has only begun to be investigated. But early reports suggest that its origins date to the dawn of the early modern era.

Authorities on this history include Jeremy Black and Walter Goffart, who have separately trained their sights on the rise of the historical atlas genre.[1] Both scholars start by observing that *all* premodern maps were saturated with history. Their evidence comes

from Europe, where the medieval T-O map serves as a shared point of reference. In its simplest form, the T-O was a timeless diagram, characterized by a circle or oval encompassing the three continents known to the ancient Greeks. In Christian hands, however, this spatial scheme was given a third dimension, as it were, being recast as the staging ground of the founding biblical drama: the expulsion of Adam and Eve from Paradise and the dispersal of their progeny across the world. Medieval Christians learned that Eden lay at the far edge of Asia, and that the sons of Noah had each headed to a different continent; the names Shem, Ham, and Japhet were emblazoned on some medieval maps as prominently as the Greek terms for Asia, Africa, and Europe.

An important feature of such maps was their temporal thickness. Just as the biblical drama was believed to continue unfolding long beyond Genesis, carrying through the transformative life of Jesus (the New Adam) and hurtling toward his promised Second Coming in the future, so any number of subsequent stories could be overlaid on the mythohistorical scheme of the T-O map. This was taken to extremes in the massive Hereford *mappamundi*. In one of many incidents crowding the map's surface, the Exodus of the Jews from Egypt is shown as a looping line crossing from Africa into Asia. Starting from the city of Ramses, the route traces the Israelites' journey north across the Red Sea to Mount Sinai, where a drawing of a stone tablet commemorates Moses receiving the Ten Commandments. A golden calf nearby signifies the point where his followers became anxious and began worshipping false gods. From here, the line twists and turns—a clever device to signal forty years of aimless wandering in the desert—before straightening out and heading across the River Jordan into the Promised Land.

In its linear and pictorial elements, this iconography would not be out of place in a modern history book. Its multilayered texture is what breaks with modern convention. At one memorable point, the Israelites' path toward the Jordan River takes them past a sketch figure commemorating an episode from generations earlier: the moment when Lot's wife was turned into a pillar of salt as punishment for looking over her shoulder at the smoldering ruins of Sodom and Gomorrah. This incident belongs not to the Exodus narrative at all, but to a much earlier story-cycle from Genesis. The simultaneity of these disparate events on the surface of the map creates a distinctly non-modern temporal jumble, where widely separated time-strata are allowed to coexist and collide.

As the reader will have noticed, this description of premodern *mappaemundi* is based entirely on European examples. East Asian materials are not introduced by either Goffart or Black, despite the implicit claims of their titles to present a universal history of the genre. This is unfortunate, for the earliest *mappaemundi* from Eurasia's far East are nearly as old as those from the far West—and they embody a sensibility much like that of their Christian counterparts. The techniques by which the Hereford map marked the journey of the Israelites out of Egypt find striking parallels in the way the oldest Buddhist world maps traced the epic journey (from China to India and back) of the Tang-era monk Xuanzang.[2] In figure

2.1, a line threading through the landscape again represents a journey—again through terrain that is thick with earlier referents (here, sites from the life of the historical Buddha, who lived 1,000 years earlier, interspersed with the names of Buddhist monasteries and kingdoms established in the intervening centuries). While modern mapmakers have created narrower slice-of-time visualizations, foregrounding the pilgrim's path against a more stripped-down backdrop, fifteenth-century mapmakers were responding to different imperatives. Working from a classic text of their own, they overlaid Xuanzang's journey onto a landscape that was crowded with toponyms from earlier times, creating a rich palimpsest for the viewer. Just as the Israelites' path toward the Jordan River took in the destruction of Sodom and Gomorrah generations before, so the magnificent Buddhist maps made in medieval Japan plotted Xuanzang's journey through a landscape flush with mementos from the previous millennium.

In short, at both ends of Eurasia, medieval *mappaemundi* (and many local maps as well) were saturated with historical content. What Black and Goffart are interested in is when—and why—Europeans began to separate historical content from general geography, gradually dividing it into shorter slices of time that could be assembled into a series in atlas form. In a word, how did the practice of a specifically *historical* cartography develop?

For Europe, the origin of this tradition is routinely traced to the Flemish cartographer Abraham Ortelius (1527–1598). In Ortelius's day, even general maps of distant places were expected to be of interest primarily to readers of literature, including the Bible and the classics; as a result, many of the maps in the *Theatrum orbis terrarum* should be read as historical reference maps. But Ortelius went a step further, including in his general atlas (starting in 1579) a separate appendix devoted to maps of the ancient world. This historical supplement, known as the *Parergon* (an archaic term meaning "supplement" or "embellishment"), grew steadily from one edition to the next until it was eventually spun off as a separate work in its own right, becoming the first dedicated historical atlas to be published in Europe.[3]

What distinguished Ortelius's historical maps from his general maps were two subtle features: their framing and their toponyms. Most maps in the *Theatrum* took political units (countries and kingdoms) as their object, whether singly or in groups. But for the *Parergon*, Ortelius downplayed geopolitics in favor of unbounded regions like the Aegean Sea or the Holy Land: arenas of action for a particular character or event. He then proceeded to fill those frames with toponyms from the Bible or classical texts, creating locational referents for the many foreign place names that European readers would encounter when reading ancient works. Thus "The Wanderings of Abraham" offered a landscape saturated with place names from the book of Genesis, while "The Realm of the Argonauts" presented a Mediterranean marked with topoi from Greek mythology.

As the *Parergon* evolved, Ortelius also found room for a second kind of historical cartography—one that has also had a lasting influence on the field. Starting in 1624, he began reproducing a facsimile

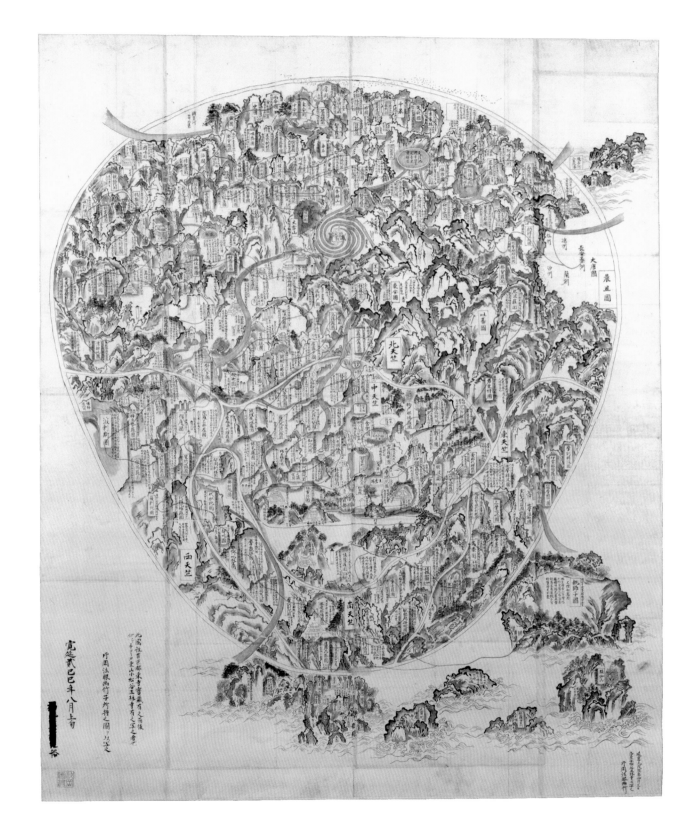

FIG. 2.1 *Map of the Five Indias* (Gotenjiku zu), Japan, 1749 (copy of 1364 original), 167 x 133 cm, Kobe City Museum.

map, the *Tabula Itineraria* of Peutinger, alongside maps of his own design. As a late medieval copy of an imperial Roman route map, the Peutinger Table presented a vision of Eurasia that was severely outdated by the seventeenth century.[4] The point of reproducing this itinerary was as an exercise in intellectual history: an attempt to grasp how people of the late Roman empire understood their own world. We have other evidence that Ortelius was intrigued by the mental maps of the ancients; for the first illustration in the *Parergon*, he created a novel image that projected the portion of the world known to Ptolemy onto a larger oval representing the globe as understood by Ortelius's contemporaries. In addition to demonstrating the limits of the ancients' geographical knowledge, Ortelius used this image to convey their climate-zone theory, making it serve as a dual vehicle for intellectual history. The Peutinger Table, with its distortions of direction and exuberant medieval iconography, was ready-made to serve a similar function.

In short, Ortelius's *Parergon* already combined two distinct modes of historical cartography: the effort to reconstruct bygone landscapes, and the reproduction of old maps as clues to the geographical imaginations of their makers. Once it was clear that there was a market for these images, many others followed in Ortelius's wake. Not surprisingly, anachronisms crept in; modern-looking ramparts might surround ancient cities, and sixteenth-century armadas might turn up in third-century seas. But whatever their stylistic shortcomings, these images had generative power. Ortelius had left Europe with a fertile seed-bed in which a vigorous historical cartography could grow and flourish.

And what, one might ask, does this have to do with Japan?

In surprisingly short order, Ortelius's fame extended to the far reaches of Eurasia. Within a decade of its publication in Antwerp, the *Theatrum* was carried to Beijing by the Jesuits, who presented a copy to the Ming emperor in 1600. Another copy made its way to Edo at about the same time. The Italian Jesuit priest Matteo Ricci (1552–1610) drew on Ortelius (among other sources) when drafting his landmark world maps in Chinese—including the famous wall-sized, woodblock-print version of 1602 that quickly found its way to the Chinese, Korean, and Japanese courts.[5] Two hundred years later, the *Parergon* figure described above—the first image illustrating the worldview of the ancient Greeks—showed up intact as a cartouche on a Japanese folding screen.[6] Yet Ortelius was in no position to serve as the progenitor of historical cartography in East Asia. Even the most literate intellectuals in the Ming, Joseon, and Tokugawa realms would have lacked the cultural literacy needed to make sense of the *Parergon*. Indeed, most readers in Asia were probably not even aware of whether they were looking at historical or general maps.

Nor was the mapping of time a major concern of the later European geographies and atlases introduced to East Asia during the early modern period. This goes for the writings of the Italian Jesuit Giulio Aleni and his fellow Society of Jesus missionaries in the seventeenth century, as well as for the German and Dutch geography texts that entered the Japanese

port city of Nagasaki in the eighteenth and nineteenth centuries. As a rule, these primers showed little interest in history; their focus was on the physical and political contours of the globe in their own day. So, to the extent that historical mapping developed in Japan during the early modern period, it cannot have been in response to European stimuli.

Yet historical mapping did indeed take off independently in early modern East Asia, flourishing in three separate domains. One branch followed an existing tradition with deep roots in Chinese statecraft, mapping the emperor's realm and its administrative apparatus. A second line of inquiry stayed closer to home, plotting traces of the past in the native-place landscapes of provincial literati. And a third genre took the form of event maps. Some of the latter declared their subject matter openly, but others were coy. It has taken scholars a long time to identify some of these as historical maps, for they masqueraded as atemporal—and thus apolitical—portraits of place.

The following essay takes up each of these three developments in turn. Its animating questions are threefold. First, under what circumstances did Japanese of the Edo period (1603–1868) turn to maps as a way to locate themselves in time as well as in space? Second, how can we characterize the resulting maps? And finally, in what ways does the East Asian archive challenge accounts of historical cartography that lean exclusively on the European case?

Mapping the Realm

Long before Ortelius, historical cartography was entrenched and thriving in East Asia as an element of statecraft. The long cycle of dynastic collapse, succession, and reunification—a process that never transpired in the lands of the former Roman empire—created the context for historical mapping of the Chinese state. This must surely be the oldest continuous tradition of explicitly historical mapping in the world; it certainly has a long and lustrous pedigree. Maps showing how previous dynasties subdivided and ruled the imperial realm (C. *tiensha*, J. *tenka*, lit. "all under heaven") can be attested for nearly a thousand years, going back at least to the Song dynasty (960–1279).[7] As early as the twelfth century—some four hundred years before Ortelius—a stone stele entitled "Map of the Tracks of Yu" (*Yu ji tu*) was erected on the grounds of a Chinese academy. Although celebrated today for its mathematical accuracy more than its historical content, this remarkable map represents an attempt to reconstruct the ancient landscape described in the *Tribute of Yu*, a mythohistorical text that was itself already centuries old by the time the stele was carved. Standing intact to this day, the granite tablet has served for centuries as a printing press of sorts; by taking rubbings on paper, generations of students have been able to produce their own reference maps to consult while poring over ancient texts.

Woodblock maps from the Song dynasty with titles that draw attention to their historical content continued to be reprinted for centuries. Claiming to illustrate "the Ancient and Present Territories of China and Foreign Countries," or to present a "General Map of Distances and Historic Capitals," images

FIG. 2.2 *Map of China* (Chūgoku zu), 1727, Terajima Ryōan, 160 x 260 cm, Kobe City Museum.

in this mold circulated throughout the Sino-sphere, serving as prototypes for a historically attuned cartography of the realm in Korea as well as in Japan.

The first printed Japanese historical maps of this kind were in fact direct copies of Chinese originals. But over time, Japanese intellectuals begin to compile and annotate their own historical maps of the realm. Figure 2.2 is a case in point. Hand-painted on silk, it was made by Terajima Ryōan (1654?–?), a doctor from Osaka who earned lasting fame as editor of the 105-volume *Illustrated Sino-Japanese Encyclopedia* (*Wakan Sansai Zue*). Like its Chinese prototype (*Sancai Tuhi*), Terajima's 1712 compendium featured a number of world maps, including a crude caricature of Matteo Ricci's massive wall map for the Ming court. Here, by contrast, the learned doctor filled a large silk canvas with painstaking detail. In addition to the scores of toponyms crowding the map proper, he supplied the names of former emperors (near their capitals) and a complete table of past capitals (in the lower right-hand corner).[8]

Given the deep roots of this kind of cartography in the Chinese tradition, it is not surprising to learn that the first full-color historical atlas to be published in Japan was a stroll through the history of China (figs. 2.3a and 2.3b).[9] The author of this landmark work, Nagakubo Sekisui (1717–1801), was as prolific a scholar as Terajima. Before publishing his woodblock historical atlas (which covered two millennia of Chinese dynasties, from the Zhou to the Qing, in thirteen sheets), Sekisui had made a splash with a gridded sheet map of Japan.[10] He had also created a massive wall map of the core territories of the historical Chinese empires. Although its title referred

Orienting the Past in Early Modern Japan   [43]

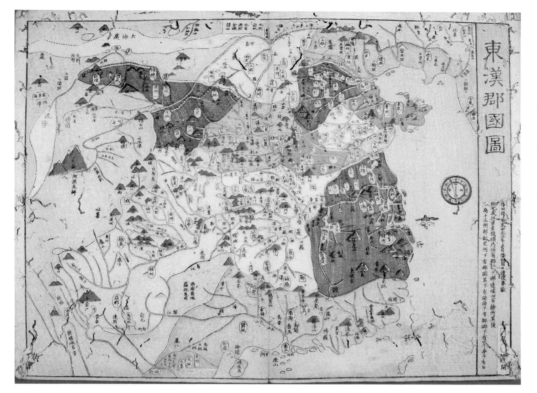

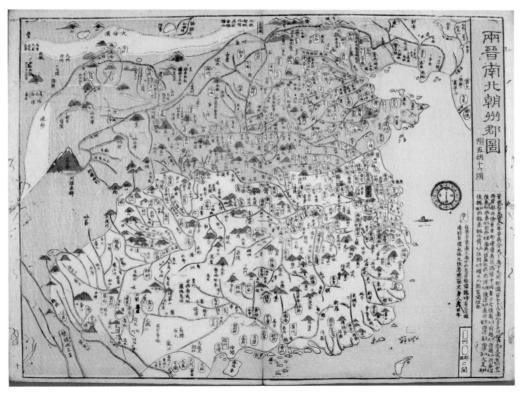

FIGS. 2.3A AND 2.3B Sample pages from *Historical atlas of Chinese provinces and districts under each successive dynasty* (Tōdo rekidai shūgun enkaku-zu), 1st ed. 1797, Nagakubo Sekisui, 36 cm (= length of the book), Waseda University Library. On the top is map 7, showing the Eastern Han, and on the bottom is map 9, Northern & Southern Song.

FIG. 2.4 *Panoramic view of the famous places of China* (Tōdo meisho no e), 1840, Katsushika Hokusai, 41.5 × 54.5 cm, Kobe City Museum.

to the reigning Qing dynasty, the recent conquests of the Manchus in the north and west were left off the map, confirming that historical interest was what drove its design. There was no need to include Mongolia, Tibet, or Xinjiang, given that Sekisui primarily intended his map to help his fellow scholars grasp the landscape of the Confucian classics.

Japan's leading cartographic historian of the twentieth century, Unno Kazutaka, has speculated that Sekisui may have incubated the idea of making a comprehensive time-map of China for nearly two decades. Unno deduced that the experience of trying to cram all the place names that were needed as reference points onto a single sheet—even one as large as two meters on a side—might have been what prompted Sekisui to try his hand at narrower time-slices in the form of a historical atlas.[11] In any case, that was exactly what he did, completing the atlas just five years after publication of the wall map. Starting with a general route map of the Qing empire, it proceeded to show the changing territorial limits of successive dynasties, the shifting location of their capitals, and the administrative units (provinces and counties) through which they carved up their realms.

Orienting the Past in Early Modern Japan [45]

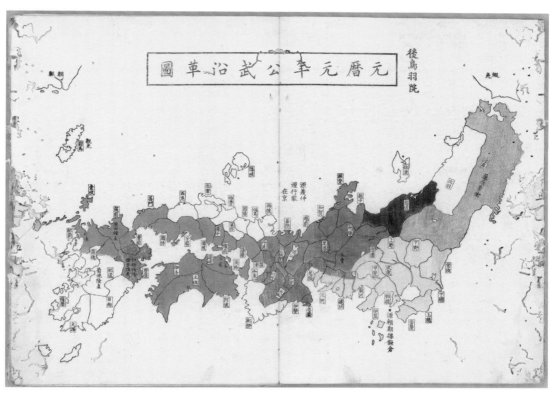

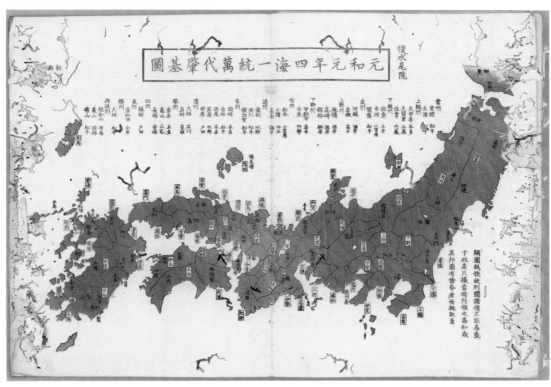

FIGS. 2.5A AND 2.5B From *Atlas of Eternal Peace* (Bansei taihei zusetsu), 1815, Hiyama Gishin, 32 cm (= length of the book). Top: *Court nobles and Samurai in 1184* (map 1); bottom: *Map of the unity of the whole country in 1615* (map 11), Waseda University Library.

Sekisui's pioneering cartography left a rich legacy for later geographers and printmakers in Japan. On the one hand, he inspired further experiments in nostalgic mapping of the Chinese past. Figure 2.4 shows one intriguing image indebted to Sekisui's vision: a panoramic view by Katsushika Hokusai (1760–1849) of China's "famous places" (*meisho*).[12] Hokusai here deployed an oblique perspective and a pictorial idiom that would have been instantly recognizable to anyone who had seen his stylish bird's-eye views of famous places in and around Japan, although this technique had never before been applied to China. The result was not a "thin" scholarly map but a "thick" and alluring image that functioned very much like the maps in Ortelius's *Parergon*. Crowded with toponyms from classical poetry and literature, Hokusai's panorama was designed to facilitate readers' imaginary journeys into a remote and distant past.

Sekisui also helped set an example for more scholarly historical atlases, including the first one to apply this technique to the history of Japan itself. The pioneering *Atlas of Eternal Peace* (figs. 2.5a and 2.5b), published in 1815, presented the country's unstable medieval geography in eleven maps, running from the contending warrior powers of the 1100s through a series of civil wars and temporary truces until it culminated in a reassuringly boring, monochromatic image of reunification under the Tokugawa shoguns. (A later volume by the same author would tackle earlier centuries, constituting a kind of prequel to the 1815 work.)[13]

In these ways, Japanese cartographers of the Edo era were able to build on a deep foundation of historical mapping of the political realm, experimenting with print forms suitable for ordinary readers as well as for scholar-officials. Even when published for the market, however, this kind of mapping retained traces of its roots in big-picture cartography, embodying a top-down vision with a strong focus on the territory of the state. Maps like this played an important role in shaping the geographical imagination of readers in early modern Japan. On the one hand, they sustained a strong sense of participation in a pan–East Asian past. On the other hand, they helped cultivate a distinctive sense of Japan's own national identity, modeled on but separate from that of its mighty continental neighbor.

Mapping the Native Place into National History

It was in this context that a second branch of historical mapping began to flourish in the Japanese countryside, focusing on beloved landscapes closer to home. This diffuse project of regional or chorographic mapping was part of a broader mid-Tokugawa turn toward recording and celebrating local history, an enterprise that left traces all over eighteenth-century Japan. During the same decades, local literati throughout the archipelago appear to have busied themselves founding village schools, composing native-place poetry, writing local histories, and erecting stone stele to commemorate their ancestors. And many of them began to draw historical maps. Typically, this native-place cartography was undertaken by relatively affluent amateurs, scions of landed families with

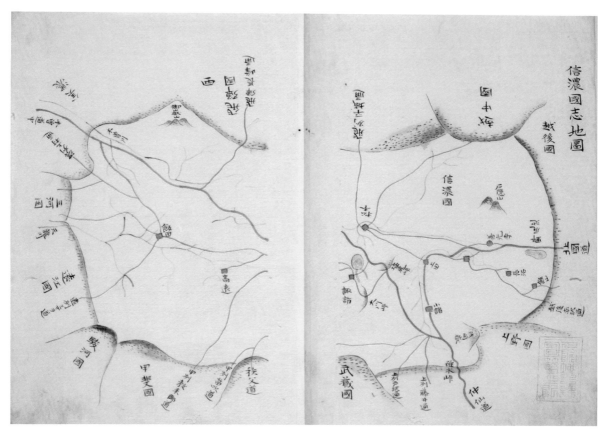

FIG. 2.6 *Sketch map of Shinano Province* (Shinano kokushi chizu), from *Shinano Kokushi* [History of Shinano Province], n.d. (18th c.?), author unknown, National Archives of Japan.

distinguished pedigrees. For such men, local cartography was intimately linked to family genealogy. Plotting the traces of ancient imperial pasturelands or medieval fortresses in their own backyard provided a way to link their personal and provincial pasts to the prestigious pageant of imperial history.

History mapping of this kind was a bottom-up enterprise, albeit one that developed within the well-established context of top-down (state-led) mapping. In China, regional cartography had been carried out repeatedly over the centuries as part of official provincial and county gazetteers, but such was not the case in Japan. Although scattered examples of this form survived from the ancient period, provincial record-keeping in the Japanese archipelago broke off almost entirely during the long centuries of civil war. Even after the unification battles of 1600, Tokugawa rule was premised on the continued parcelization of sovereignty among some 260 daimyo domains (whose territories overwrote the ancient provincial order). Lacking the power to revive the centralized imperial administrative apparatus, the shogunate left local affairs—and meso-level historical cartography—almost entirely in the hands of local enthusiasts operating outside the context of state support. To be sure, the men who created these maps often gained cartographic training in the course of service to the regime. But beyond that, early modern Japanese who wanted to plot the local past were pretty much on their own.

The result was an exuberant, bottom-up culture of local historical cartography that developed along two major pathways. One stream consisted of amateur maps, made for personal pleasure and edification, which circulated among friends in manuscript form.[14] Figure 2.6 shows one such map of Shinano Province, made by a history buff in the foothills of the Japanese Alps. In Shinano, local historians were often keen to identify the locations of ancient imperial pasturelands—important sites for a region that supplied the court with horses—as well as the ruins of medieval fortifications, former post-stations, and toponyms mentioned in classic chronicles or literary texts. As noted above, the research required to recover these traces of the national past in a given locality were one expression of a general boom in local history.[15]

If amateur history buffs created one kind of local history-mapping, however, another kind arose in response to the market. In a society marked by rapid urban growth, rising literacy rates, and a flourishing print market, publishers found a growing constituency for woodblock prints of many kinds. Among the many single-sheet prints that found their way to the bookstalls of Edo—alongside images of scenic spots and sumo wrestlers, actors and geisha—were imaginative reconstructions of past landscapes. Not just any locale merited such treatment. Judging from the surviving examples, the primary force behind history-mapping for the market was travel. It makes sense that pilgrimage destinations and major cities were the most common subjects; both commanded a sizeable, moneyed constituency, including thousands of tourists for whom a printed map represented an appealingly low-cost, lightweight souvenir.

Figure 2.7 pairs two examples of this colorful genre. The image on the top (fig. 2.7a) conjures the lost landscape of Hiraizumi, a veritable ghost town in northern Honshu. During the twilight years of the classical era, Hiraizumi had flourished as the headquarters of the northern Fujiwara clan until it was destroyed in 1189. By the eighteenth century, when this map was made, only a scattering of old temples marked the spot, presiding over a desolate sea of grass. (The poet Bashō [1644–1694], who trekked through the area a hundred years earlier, was moved to sigh, "all that remains of the warriors' dreams is summer grasses.") Clearly, the open landscape that a visitor would have seen while walking through this terrain in the eighteenth century would have differed radically from the medieval town conjured by the map, with its gardened temple compounds and well-ordered commoner districts crowded cheek-by-jowl along the river. In this case, a history-map served as a tool for imagining a bustling bygone world.[16]

Equally remote from contemporary experience—though in a different direction—is figure 2.7b, an attempt by the talented Japanese artist Utagawa Sadahide (1807–1878) to evoke the marshy landscape of Edo Bay before it became the headquarters of the Tokugawa shogunate.[17] When this image was printed in 1853, Edo was a massive metropolis; to house its million residents, the waterfront had been extensively filled in and built up for two centuries and more. As a result, this unimposing image represents an impressive exercise of the imagination: an attempt to

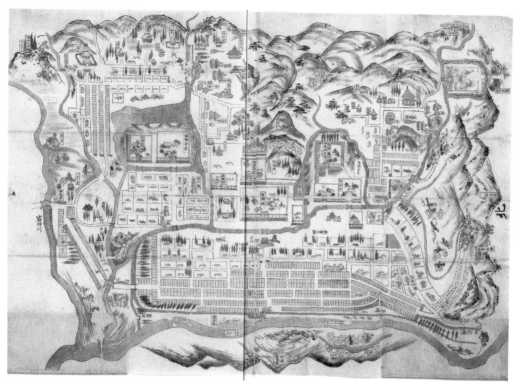

FIGS. 2.7A AND 2.7B Top: *Historical map of the abandoned city of Hiraizumi*, n.d. (18th c.?), author unknown. Reproduced in Yamashita Kazumasa, *Edo-jidai kochizu o meguru*, 202–3, ©Yamashita Kazumasa; bottom: *Historical map of Edo Bay*, from *Oedo zusetsu shūran* [Collection of old Edo maps], 1853, Hashimoto Kenjirō (Utagawa Sadahide), Kobe City Museum.

wind back the clock 250 years, to suggest what the marshlands around the bay might have looked like when the human footprint was limited to a dozen small fishing villages. That a print like this would be published at all suggests the Japanese public's growing fascination with geographical metamorphosis at a moment when even more rapid change was about to set in.

Edo may have been the largest metropolis in early modern Japan, but it was not the chief object of this kind of nostalgia. It was Osaka, the realm's second city and a major center of merchant culture, that gave rise to more historical maps than any other in Japan in the early modern period. In fact, this genre had its own label, derived from an earlier name for the port: "Naniwa old maps." (Some entries in this category masqueraded as copies of medieval maps, but most if not all were actually made in the eighteenth or nineteenth century.) As was true of Edo, reclamation had pushed the waterfront out dramatically, so that for viewers familiar with the many reference maps of the contemporary city that circulated in their own day, the shoreline on these "old maps" would have immediately registered as a sign of times past.

Second only to Osaka as an object of loving reconstruction was the venerable imperial capital of Kyoto (Heian-kyō). Modeled on the great T'ang capital of Chang-an when first built in the late 700s, Kyoto was ravaged repeatedly by fire and war over the years. Although it remained home to the imperial household for a millennium, its geography morphed almost beyond recognition. This created plenty to exercise the imaginations of historians. While the Edo-era city retained the grid pattern of its original streets, a yawning gap separated the eighteenth century from the eighth century, and local literati were eager to fill it in. Of the three maps comprising figure 2.8, the image on the left (fig. 2.8a) reflects an Edo-era vision of the original city plan from the late 700s. Next to it (fig. 2.8b) is a recreation of the burned-over medieval city in the 1400s, its urban fabric ravaged by civil war, while the final image in the series (fig. 2.8c) imagines the city two centuries later, when its dominant visual feature was a defensive wall around its shrunken perimeter.

The man who created these time-slice maps of Kyoto was one Mori Yukiyasu (also known as Mori Kōan, b. 1701), one of the most prolific and ambitious historical cartographers of his day. In the course of his career, Mori drew a remarkable 400 maps. Some were tracings from borrowed originals; most were intended for inclusion in a never-completed national gazetteer. This ambitious volume would have framed Japan at a succession of scales, starting with its location in the world before focusing on each province in turn. But the area on which he planned to lavish the most attention was the capital region. In addition to the three images of Kyoto shown here, Mori archived many more, including a meticulous block-by-block reconstruction of the capital during the classical Heian period.

One of Japan's leading map historians today, Uesugi Kazuhiro, has devoted much of his career to studying this remarkable cartographer. In the process of probing Mori's archive, Uesugi also articulates a manifesto for a new kind of research in historical car-

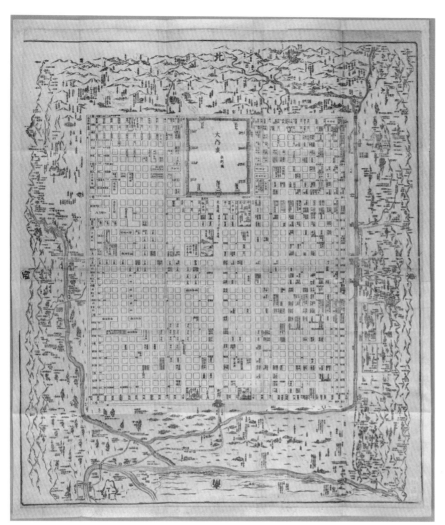 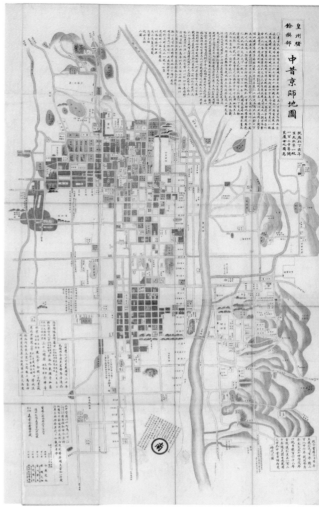

tography. Where most Japanese map historians have focused on the "vertical" transmission of elements from one map to another over time, Uesugi calls for "horizontal" analysis, deemphasizing sources and legacies in favor of reconstructing the social and material worlds in which these artifacts were made, seen, and shared. Accordingly, Uesugi has diagrammed the networks of map borrowing and lending that men like Mori were enmeshed in—networks that made their work not only influential but conceivable in the first place. (No single collector, he points out, could have owned prototypes of the hundreds of maps Mori traced. He could only have amassed those originals through extensive borrowing.)[18] Uesugi's diagrams look much like the digital images being produced by the Republic of Letters project, where scholars are collaboratively mapping out the correspondence of Enlightenment figures such as Benjamin Franklin or Athanasius Kircher.[19] As these scholars have brought to light a transatlantic republic of letters, so Uesugi has diagrammed a Japanese "republic of maps."

The most interesting figure to turn up in Uesugi's study is the founding father of Japanese nativism,

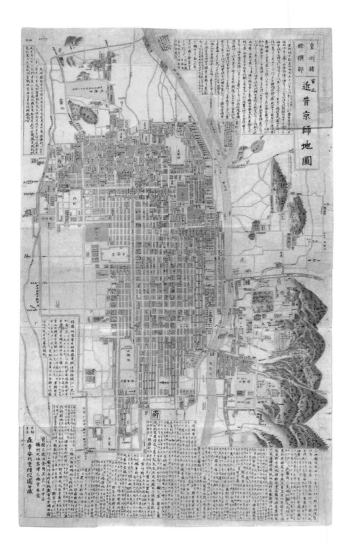

FIGS. 2.8A, 2.8B, AND 2.8C Three Kyoto historical maps. Left: *End of Heian-Kamakura* (Karaku ōko zu), 1791, author unknown, 93 × 80 cm, National Diet Library; middle: *1467–1592* (Chūjaku Keishi chizu), 1753, Mori Yukiyasu, 121 x 77 cm, National Archives of Japan; right: *17c* (Kansei kinjaku Keishi chizu), 1753, Mori Yukiyasu, 128.0 x 76.0 cm, National Archives of Japan.

Motoori Norinaga (1730–1801). Famous among intellectual historians for his philological and ideological writings, Norinaga turns out to have been an avid maker and collector of maps as well. Careful sketch maps (principally of Kyoto) made during his student years are still preserved in the Motoori archive. Later, he set his son to the same task; of the eighty-odd maps preserved in the family storehouse, the majority were made by Norinaga's son, probably as part of his homeschooling.

Most of this cartographic archive consists of copies based on someone else's original (whether print-ed or hand-drawn). But one telling exception—and a standout in the collection—is a hand-drawn black-and-white map identified only as "Castle town of the Hashihara clan." As it happens, the Hashihara clan did not exist; the young Norinaga dreamed them up. At the age of nineteen, he drew a genealogical chart for the fictive clan and designed an imaginary city for them to preside over, giving it a gridded structure reminiscent of Kyoto. If this exercise has an echo in Europe, it may well be the imaginary map that accompanied Thomas More's *Utopia* (1516). Like More, Norinaga was a utopian thinker. He spent his

Orienting the Past in Early Modern Japan [53]

life sifting the Japanese annals for their earliest elements, calling on his countrymen to rid their culture of every foreign import, and dreaming of a day when the Japanese would return to their pure original essence. The maps he made as a teenager show that, from the beginning, Norinaga grounded that dream in a mythical past landscape—one that both was and was not the historical capital of Japan. In a fascinating twist, this utopian thinker turned historical cartography to account as a blueprint for the future.

Mapping the News

So far we have looked at two broad branches of historical mapping in East Asia. The older of the two, concerned with the administrative geography of the imperial state, was fundamentally a top-down enterprise; the newer one, concerned with local pasts, was essentially a bottom-up effort (whether undertaken by amateur historians or professional artists). But by the end of the Edo era, a third branch had appeared on the time-map tree as well: the cartography of current events.

Censorship was the crucible in which this enterprise evolved. In Japan under the shoguns, commoners were not allowed to speak about public affairs; there was no sanctioned medium for broadcasting news. Nonetheless, rumors piqued people's curiosity, and publishers learned that if they moved quickly enough, they could make a profit from informative broadsheets known as *kawaraban*. To foil the censors, these handbills were often produced without dates or attribution; both the designer and the publishing house typically remained anonymous. Because events unfolded in particular places, the *kawaraban* often included maps. In particularly sensitive cases, the map itself might stand in for the news.

The most common theme for *kawaraban* cartography was natural disasters. In a highly urbanized country where cities were made almost entirely of wood, it is no surprise to learn that one common calamity (and one that lent itself to sensational imagery) was urban fire.[20] But floods were common as well, and could damage vaster regions. The catastrophe behind figure 2.9 was a particularly disastrous event from 1847, precipitated by an earthquake and heavy rains in the Japanese Alps. Published in Edo a few months later, this anonymous broadsheet combined a multicolor map with a discursive account of how the disaster unfolded. As the text explains, the incident began when an earthquake set off a mudslide that blocked the Sai River. Unable to drain out of the valley, rainwater backed up above the slide until it covered a score of villages (whose individual names are written in small oval cartouches in the upper left, inside a swollen band of dark blue ink). After three weeks, the water massed behind the temporary dam burst through, flooding dozens more villages in the Zenkōji plain downstream. The villages affected in the second stage are again identified by name, their cartouches crowding the light blue lake that covers the heart of the page. The resulting image is one of a handful from this period that pushed the envelope in terms of technique, using differential shading to convey sequential stages of a long-running, multipart event.[21]

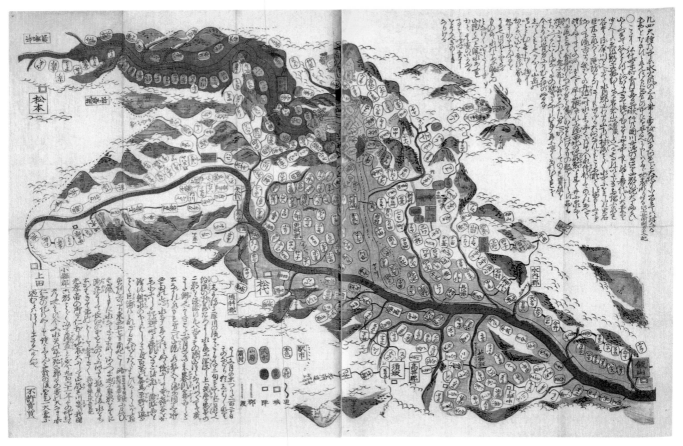

FIGURE 2.9 *Map of the great disaster in Shinano Province* (Shinshū saigai no zu), 1847, author unknown, publisher unknown. Reproduced in Yamashita Kazumasa, *Edojidai kochizu o meguru*, 194–95, ©Yamashita Kazumasa.

If publishing about natural disasters required evasive action, even more care was required when reporting on protests, uprisings, or battles. Yet these, too, occasionally found their way into print. In fact, the oldest *kawaraban* that survives today—an outlier from the seventeenth century—depicts the siege of Osaka Castle in 1615.[22] It is not clear how much time had elapsed after the battle before this print appeared on the market, but even if it came out quickly, this was a relatively safe subject; the victors in the battle were firmly ensconced as the recognized rulers of Japan at the time it was made. And it was certainly not risky in the nineteenth century to depict the famous medieval battles of Kawanakajima (where the armies of Takeda Shingen and Uesugi Kenshin faced off repeatedly for a decade), which the Japanese public could view at the comfortable remove of 300 years. Current events in the nineteenth century were another matter altogether, however—and one that called for a more subtle approach.

The difference is dramatized by two contrasting prints of volcanic Mount Asama, produced seventy-five years apart. One was a standard anonymous broadsheet illustrating a terrible disaster that unfolded after the volcano erupted in 1783. In an eerie echo of the flood described above, a massive lava flow blocked the upper Chikuma River, causing extensive flooding downstream when the dam eventu-

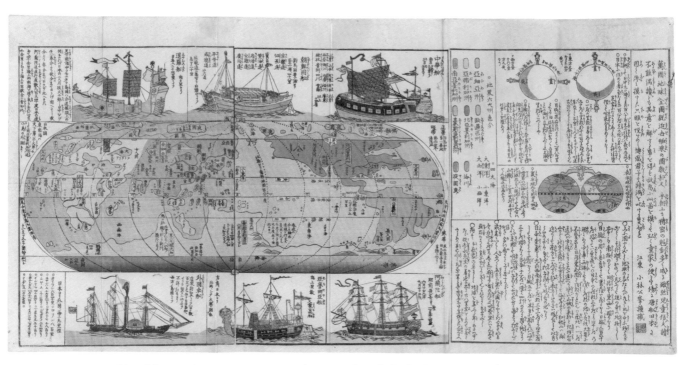

FIG. 2.10 *Map of the myriad countries, mountains, and seas, at a glance* (Sekai bankoku zenzusetsu), Edo era, Kobayashi Kōhō, 40 × 77 cm, Yokohama City University Library.

ally burst (a scenario that was all too familiar for the beleaguered residents of northern Shinano). A later print of the same mountain, by contrast, invites a different reading. Made by the *ukiyoe* master Utagawa Hiroshige (1826–1869), it appeared on the market as part of the series *One Hundred Famous Views in the Various Provinces* in 1859. At the time, Asama had not erupted for thirty years. Interestingly, the same artist had made prints with the mountain in the background before, but never had it loomed in such a menacing, ominous way as it did in his 1859 print. Given the chaos that was beginning to engulf the shogunate by then, it has been suggested that Hiroshige may have been treating the volcano as an allegory for the political rumbling and smoldering that was going on just under the surface of Japanese society in the late 1850s.[23]

Hiroshige certainly would have had reason to be concerned. Since the turn of the century, British and Russian ships had been making incursions into Japanese waters with increasing frequency. Just six years earlier, Commodore Matthew Perry's "black ships" had steamed into Edo Bay to demand access to Japanese ports. These events prompted a flurry of handbills that caricatured the ships, conveyed details about the barbarians' demands, and mapped the shogun's response. If some of these *kawaraban* were overt about their subject—bearing bold titles like "Perry arrives in Uraga," or "Amassing of forces for the defense of the divine land"[24]—others were more coy. Kobayashi Kōhō's "Map of the myriad countries, mountains and seas, at a glance" (fig. 2.10) is a case in point. A censor who saw this innocuous title in a catalogue might assume it was a straightforward world map. Yet the crude Ricci-style map at the heart of the image is visually overpowered by six foreign vessels along its margins—including one (bottom center) labeled "Russian ship, a.k.a. Damn Mongol ship."[25]

[56] CHAPTER TWO

FIG. 2.11 *Map of the boundary between Hitachi and Shimōsa Provinces* (Jōsō kyōkai no zu), 1865, author unknown, Ibaraki University Library.

As Japan entered the decade of the 1860s, publishers had to tread ever more carefully. Fear of the foreigners prompted a firestorm of angry reaction from Imperial loyalists against the shogun, continuing for a dozen years after the Tokugawa acceded to Perry's demands. Incidents of terror and eventually open rebellion against the government made for sensational news. But how was a gagged press to handle that? Once again, maps were pressed into service.

The lengths to which cartographers might go to get around government censors are illustrated by an image dating from the year of the so-called Tengu Insurrection (Tengutō no ran). In this armed clash of 1865, a small band of warriors from Mito domain marched to the top of Mount Tsukuba, north of Edo, and proclaimed a new government in the name of the emperor. Their plan was for the Lord of Mito to join them and lead a march on the capital, but that did not happen. Instead, the small band of twenty-four Imperial loyalists—and more than 1,000 followers—were defeated; 353 were executed, and many more slaughtered on the battlefield. It is easy to imagine what a sensation this caused—and how avid the public would have been to read about it. Yet directly reporting on these events was not allowed. At the time, the best a Japanese publisher could offer was a map of the site where the battle had taken place (fig. 2.11).[26] The innocuous title under which it was sold—namely, "Map of the boundary between Hitachi and Shimōsa Provinces"—is flagrantly misleading, since the boundary is nowhere to be seen. Instead, the most conspicuous feature of the map is the steep flank of

Orienting the Past in Early Modern Japan [57]

the mountain itself with its shrine midway up the slope and a suggestion of the long staircase climbing to the peak—site of the rebels' last stand.

A colleague and I have recently made a similar case for another print, issued in the same year under the innocuous title "Fifteen provinces of the Northeast." In this case, a preexisting tourist map—complete with a pop-up Mount Fuji and enlarged insets of such popular destinations as Kamakura and Yokohama—was hastily repurposed as a news map simply by tacking on a schematic rendering of the northeastern provinces.[27] The word *Tōhoku* (northeast) in the title has led more than one contemporary viewer to be disappointed, since in fact the northeast gets the sketchiest treatment of any region on the map. But in a year when ground battles were raging in that region, such a title might well have prompted sales. Maps could gesture toward what texts could not: the unfolding of momentous historical events in the archipelago.

The importance of reading innocent-looking panoramas and maps from this era as politically charged documents has been brilliantly articulated by University of Tokyo historian Sugimoto Fumiko. Sugimoto drives her point home with a close reading of two works from the prolific *ukioe* artist Utagawa Sadahide, who published a raft of geographical images in the late 1860s. Among them are a striking map and panorama of northeastern Honshu—an area that had not previously been an attractive subject for commercial views, but that became suddenly notorious during the late 1860s as the site of fierce opposition to the new regime. Both the regional map and the pictorial panorama went on the market in 1868, the fateful year when the Tokugawa government fell. As Sugimoto points out in illuminating detail, Sadahide's composition for the bird's-eye view highlighted the key regional players and battlegrounds in the Restoration wars, foregrounding the Tokugawa loyalist holdout of Aizu-Wakamatsu. Without calling attention (in his title) to what he was doing, she is convinced, Sadahide was surreptitiously publishing event maps.[28]

If Sugimoto is right, her findings suggest that we should also revisit the crude world maps that flooded the Edo market in the mid-nineteenth century. These too could serve as event maps—indeed, for our purposes, as historical maps. Attending closely to their titles sometimes turns up surprising clues. Figure 2.12 offers an instructive example. One of many crude, outdated maps in the mold of Matteo Ricci that flooded the market in these years, this anonymous broadsheet was listed under the typically bland heading "Distances by sea from Japan to the myriad countries of the world." But in the bookstalls of Edo, it could also be displayed backward, to show the more pointed title printed on the reverse: "Map of the myriad countries to quickly grasp the expulsion of the Mongols" (*Mōko taiji bankoku hayawakari zu*). For at least some mapmakers, in other words, the West's incursions into the nineteenth-century Pacific were seen as a second coming of the Mongols. The same striking historical allegory made its way into the fine print of the world map reproduced as figure 2.10 (above), where the Russian ship was marked "also known as a Damn Mongol ship

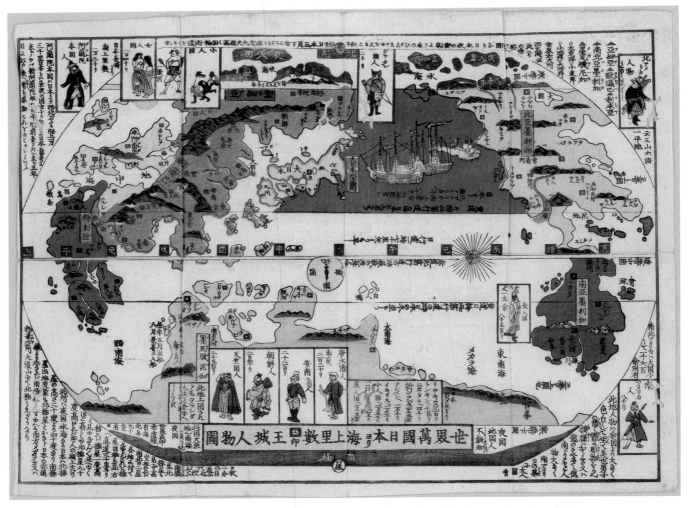

FIG. 2.12 *Map of all the countries of the world and pictures of the peoples, showing the capitals and the distances from Japan* (Sekai bankoku Nihon yori kaijō risū ōjō jinbutsu zu), a.k.a. *Map for quickly grasping the expulsion of the Mongols* (Mōko taiji bankoku hayawakari zu), ca. 1853, 35 x 46 cm, Yokohama City University Library.

(*Mōko mebune*)." In a starkly censored public sphere, maps were one of the few media through which Japanese readers could grapple with what that analogy might mean.

Although these kinds of images have not generally been treated under the rubric of historical maps, I concur with Sugimoto Fumiko that it is important to begin recognizing them as such. Japanese men and women who lived through the upheavals at the end of the Tokugawa shogunate experienced their times as epochal. Where their forebears had developed time-maps to reconstruct nostalgic landscapes of the past, Japanese of the 1860s turned to cartography to locate themselves in a fast-changing present. In a period of pervasive uncertainty and upheaval, with the only order they had ever known collapsing around them, educated Japanese sensed that they were living through a historical watershed. And they turned to cartography for orientation in time as well as in space.

Orienting the Past in Early Modern Japan

FIG. 2.13 *Complete Picture of the Newly Opened Port of Yokohama* (Gokaikō Yokohama no zenzu), 1859, Utagawa Sadahide, eight sheets joined and folded: 69.5 x 191.1 cm, National Diet Library.

No one embodied this sensibility more fully than Utagawa Sadahide, an artist whose work has figured prominently in this essay. As noted above, Sadahide's earliest map—the first of many—was a historical reconstruction of the Edo marshland (fig. 2.7b). Just a few years later, he was busy mapping the northeast, and then he embarked on a remarkable series of images depicting the boomtown of Yokohama, critical contact zone between Japan and the West. It seems only fitting to conclude this chapter by quoting at length from the inscription on Sadahide's magnificent Yokohama panorama of 1859 (fig. 2.13), which gives as good an account as any of why we should view artifacts like these as urgent historical maps:

> Once I wandered around the border area and asked where the trading firms were. No one knew where they were, nor was there a map to show the location. Therefore, I silently lamented. Now the opening of the port has been settled and the Five Nations [United States, Russia, France, Great Britain, Netherlands] have gathered here. If the conditions of the past had to be investigated to determine whether they had been profitable to our country, there would be no means to verify it. I could do nothing but regret this situation. Later, the publisher Hōzendō made a map of Yokohama. It was shown to me, and I saw the landscape, public buildings, Western-style houses, and urban buildings expanding in all directions. Now we see a map very clearly, so we can imagine the past phenomena of this area. A landscape painting is only for poets. Therefore, I encourage people to publish maps. If people want to see the scenery of this area, they can see them by means of this map, and they will still be able to see them one hundred years from now.[29]

Notes

Note: All websites accessed July 8, 2019.

1. Walter Goffart, *Historical Atlases: The First Three Hundred Years, 1570–1870* (Chicago: University of Chicago Press, 2003); Jeremy Black, *Maps and History: Constructing Images of the Past* (New Haven, CT: Yale University Press, 1997).

2. Map of Nansenbushu (Jambudvipa), late Edo-period copy of a fourteenth-century original. National Museum of Japanese History (Rekihaku) Collection.

3. Paul Binding, *Imagined Corners: Exploring the World's*

First Atlas (London: Review, 2003), 282–87.

4. For recent studies of this landmark map, probing its ancient sources and medieval context respectively, see Richard J. A. Talbert, *Rome's World: The Peutinger Map Reconsidered* (Cambridge: Cambridge University Press, 2010); and Emily Albu, *The Medieval Peutinger Map: Imperial Roman Revival in a German Empire* (Cambridge: Cambridge University Press, 2014).

5. Kunyu Wanguo Quantu, *A Map of the Myriad Countries of the World*, by Matteo Ricci. Of four surviving copies, the one owned by the James Ford Bell Foundation was lent to the Library of Congress for scanning, and is available in high resolution on the web. The left half, showing East Asia, can be accessed online at https://www.loc.gov/resource/g3200.exo00006Za/.

6. Folding screen with Japanese transcription of Joan Blaeu's world map of 1648. Kobe City Museum of Nanban Art; dated 1775. Reproduced as map 72 in Nakamura Hiraku, *Nihon kochizu taisei* (Tokyo: Kōdansha, 1972).

7. The following account relies on Alexander Akin, "Printed Maps in Late Ming Publishing Culture: A Trans-Regional Perspective," PhD diss., Harvard University, 2009.

8. Terajima Ryōan, "Map of China Through the Ages," 1727. Kobe City Museum.

9. Nagakubo Sekisui, "Tōdo rekidai shūgun enkaku chizu," first edition 1797. The Waseda University copy (a reprint from 1857) is accessible at http://archive.wul.waseda.ac.jp/kosho/rio8/rio8_01371/rio8_01371.html.

10. Nagakubo Sekisui, "Kaisei Nihon yochi rotei zenzu," first edition 1771.

11. Unno Kazutaka, "Nagakubo Sekisui no Shina zu to sono hankyō" [Maps of China by Nagakubo Sekisui and their influence], *Tōyō chirigakushi kenkyū 2, Nihon hen* (Osaka: Seibundō Shuppan, 2005), 522–48.

12. Katsushika Hokusai, "Morokoshi meisho no e," 1840. Reproduced in Kobe City Museum, ed., *Ezu to fūkei—e no yōna chizu, chizu no yōna e* (Kobe: Kobe City Museum, 2000), 55.

13. Hiyama Gishin, *Bansei taihei zusetsu* (1815) and *Honchō kokugun kenchi zusetsu* (1823). Both atlases are in the University of British Columbia Beans Collection, which can be accessed online at http://digitalcollections.library.ubc.ca/cdm/compoundobject/collection/tokugawa/id/14228/rec/1.

14. Many of these amateur maps have come down to us without the names of their creators. For a more extended discussion, see Kären Wigen, *A Malleable Map: Geographies of Restoration in Nineteenth-Century Japan* (Berkeley: University of California Press, 2012).

15. Mizumoto Kunihiko, *Ezu to keikan no kinsei* (Tokyo: Kashiwa Shobo, 2002).

16. Yamashita Kazumasa, *Edo jidai kochizu o meguru* (Tokyo: NTT Shuppan, 1996), 202–3. See also Yamashita Kazumasa, *Chizu de yomu Edo jidai* (Tokyo: Kashiwa Shobō, 1998).

17. Utagawa Sadahide (also published as Hashimoto Kenjirō) and Yoshinari Yamazaki, *Ōedo zusetsu shūran* (Edo: Eikyūdō Yamamoto Heikichi Kanpon, 1853). Image reproduced in Kobe City Museum, ed., *Ezu to fūkei—e no yōna chizu, chizu no yōna e* (2000), 56.

18. Uesugi Kazuhiro. "Mori Yasuyuki no chishi to Kyōto rekishi chizu." In Kinda Akihiro, ed., *Heiankyō—Kyōto: toshizu to toshi kōzō* (Kyoto: Kyōto Daigaku Gakujutsu Shuppankai, 2007), 99–120.

19. http://republicofletters.stanford.edu/casestudies/index.html.

20. For examples, see Yoshihara Ken'ichirō et al., eds., *Fukugen Edo jōhō chizu* (Tokyo: Asahi Shinbunsha, 1994). For a digitized selection of Stanford University's Early Modern News Sheets (*kawaraban*), see https://searchworks.stanford.edu/catalog?f%5Bcollection%5D%5B%5D=vh65obb3062.

21. "Shinshū saigai no zu," 1847. Artist unknown, publisher unknown. Reproduced in Yamashita, *Edo jidai kochizu o meguru*, 194–95.

22. This black-and-white image is reproduced in the Wikipedia article on *kawaraban*.

23. Utagawa Hiroshige II, "Shinshū Asama-yama shinkei," 1859. From the series "One Hundred Famous Views in the Various Provinces" (Shokoku meisho hyakkei). Museum of Fine Arts, Boston, accessible at https://ukiyo-e.org/image/mfa/sc232267.

24. These examples come from the Library of Congress collection of Japanese prints, which is accessible online at https://www.loc.gov/pictures/search/?sp=2&co=jpd&st=grid. The Japanese titles are, respectively, "Kita-Amerika gasshūkoku: Peruri to yū mono Sōshū Uraga ni torai su" and "Shinkoku fukui butoku anmin, Okatame taihei kagami: Izu, Sagami, Musashi, Awa, Kazusa, Shimōsa."

25. For further discussion, see Nakamura, *Nihon kochizu taisei*.

26. Yamashita, *Edo jidai kochizu o meguru*, 198. Yamashita gives the map the more descriptive, tell-all title "Map of the rebel army, Mt. Tsukuba."

27. Sayoko Sakakibara and Kären Wigen, "A Travel Map Adjusted to Urgent Circumstances," in Kären Wigen, Sugimoto Fumiko, and Cary Karakas, eds., *Cartographic Japan* (Chicago: University of Chicago Press, 2016), 112–15.

28. Sugimoto Fumiko, "Shifting Perspectives on the Shogunate's Last Years: Gountei Sadahide's Bird's-Eye View Landscape Prints," *Monumenta Nipponica* 72, no. 1 (2017): 1–30. Sadahide's panorama, entitled "Mutsu and Dewa at a glance" (Ōshū ichiran no zu), is reproduced and annotated in Sugimoto Fumiko et al., eds., *Ezugaku Nyūmon* (Tokyo: Tokyo University Press, 2011), 89. For the same artist's aerial view of this contested region, published in the same year, see "Mutsu Dewa Kokugun kōtei zenzu," in the UC Berkeley Mitsui collection (accessible online through www.davidrumsey.com/japan).

29. Image and transcription from the MIT "Visualizing Cultures" website; see https://ocw.mit.edu/ans7870/21f/21f.027/yokohama/gallery/pages/Y0044_YokohamaPort.htm.

3

Richard A. Pegg

Jesuit Maps in China and Korea

Connecting the Past to the Present

In 1794, the Chinese scholar Zhuang Tingfu (1728–1800) made a map presentation with a long title: *The great Qing dynasty world map of tribute-bearing countries with spherical coordinates, then and now* (Daqing tong zhigong wanguo jingwei diqiushi fangyu gujin tu) (fig. 3.1).[1] Three seemingly unrelated maps are displayed together. What is helpful in understanding the juxtaposition are the final three characters of the ambitious title: *gujin tu*. In English, they roughly translate as "a map of then and now." If the now was the expansive Qing empire, seen in the larger bottom map, what was the then? The answer lies in the other two smaller maps at the top corners, which point to events that occurred two centuries before, during the late Ming dynasty, when Jesuits from Europe arrived in China with radically new and unsettling methods of viewing the world and China's place in it cartographically. This essay shows how Zhuang's map and others in the late eighteenth century turned to the Jesuits' maps of the Ming past to make sense of another, modern wave of foreign interventions in Asia. Through the framework of the concept of *gujin*—then and now—a troubling "now" could be understood and perhaps managed through an encounter with an instructive "then."

The practice of juxtaposing old and new maps to teach geopolitical lessons had deep roots in Chinese scholarly approaches to the past. These were markedly different from those animating eighteenth-century Europeans, who imagined linear time and Cartesian space as objective features of the universe, transcending subjective experience and subject to precise scientific measurement. By contrast, Chinese intellectuals of the same era emphasized the role of human memory and the individual's active participation in the recovery of a past. They frequently invoked the term *gujin* 古今. By *jin* they meant the here-now; by *gu* they meant an unspecified there-then. *Gu* is often mistranslated in the binomial context as "old" or "ancient," which implies a hard separation between past and present. While this meaning may be accurate when the character stands alone, the two terms knit together—*gujin*—suggest a conversation between temporalities, a notion of time in which an unspecified past of memory is actively integrated with the present of lived experience. From their position in the now (*jin*), Chinese intellectuals could rummage through the unsorted, nonlinear past (*gu*) for patterns and precedents that would clarify the present.[2] In the notion of *gujin*, Chinese scholars actively

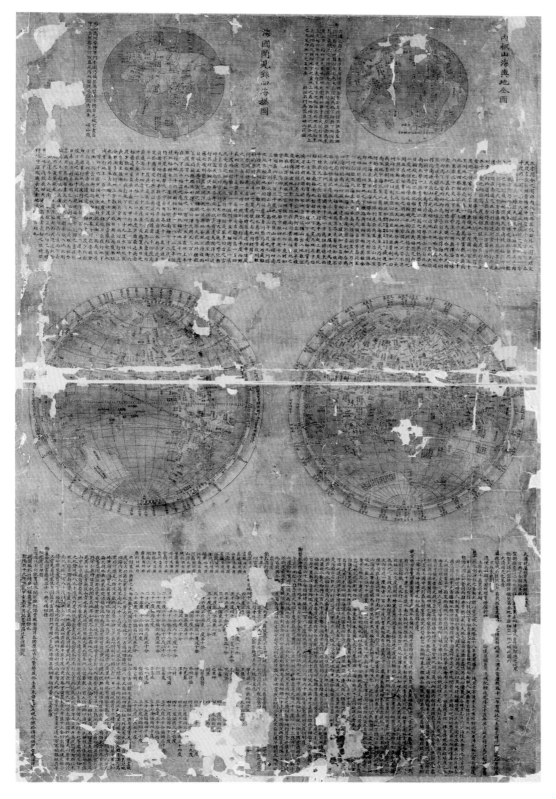

FIG. 3.1 *The great Qing dynasty world map of tribute bearing countries with spherical coordinates, then and now* (Daqing tong zhigong wanguo jingwei diqiushi fangyu gujin tu), China, Qing dynasty, Qianlong era, dated 1794. Zhuang Tingfu (1728–1800), hanging scroll, ink and color on paper, 147 x 105 cm, Library of Congress.

joined past and present through a process of deliberate recovery and application.

Zhuang Tingfu's "then and now" manuscript map of 1794 exemplifies this approach. He created it while working in the imperial court in Beijing, the symbolic center of the territorially expansive, multicultural Qing dynasty. The lightly colored medium-sized wall map, originally in the hanging scroll format, combines blocks of text with three separate maps. At the top sit two small maps that refer to the *gu*: here, the late Ming dynasty (1368–1644). Below lies a large double-hemisphere world map that refers to the *jin*: contemporary maps of the 1790s. The term *gujin* at the end of the title encourages the active association between past and present, the here-now and there-then. Zhuang's map appeared just at the moment when the West—with its advanced science and challenging worldview—had returned to China. As in the early seventeenth century, visiting Europeans had presented the imperial court with a new map of the world; and once again, the Chinese would need to come to terms with a largely unknown and potentially threatening world. When the West returned to demand opportunities for trade and diplomacy in the late eighteenth century, these late Ming maps reappeared to evoke memory associations. These juxtapositions were intended to remind viewers that this was not the first time that notions of identity were being challenged by the West, that the experience had already been processed before, and that it was time to revisit the question of China's location in the larger world.

Jesuit Maps in the Ming Court

In order to understand these Qing maps of the 1790s, we must turn back to the sixteenth century, the first moment of contact with Europe. Toward the end of the Ming dynasty (1368–1644), Jesuits bearing mechanical clocks and world maps introduced radically new notions of time and space to the Chinese court. Most famous among the Jesuit maps produced in China was one entitled *Complete map of ten thousand countries of the world* (Kunyu wanguo quantu) (fig. 3.2). This large woodblock-printed wall map was based on the geography of the sixteenth-century Flemish cartographer Abraham Ortelius but reframed around the Pacific Ocean instead of the Atlantic Ocean. Sponsored by Li Zhizao and printed on paper in six hanging scrolls by Zhang Wentao of Hangzhou, it was presented to the court by the Italian Jesuit priest Matteo Ricci in 1602. Ricci's *Complete Map* is a monument to the first cultural encounter between Catholic Europe and imperial China.[3] While the central map was crowded with over a thousand toponyms (including many foreign names transliterated for the first time into phonetic Chinese), its margins explained the movements of the sun, moon, and planets.[4]

Jesuit maps such as Ricci's challenged Ming notions of cosmological and terrestrial space. The earth-construct of most premodern Chinese maps had been essentially flat, featuring a central inhabited continent surrounded by four oceans, part of a cosmology known as "round heaven, square earth"

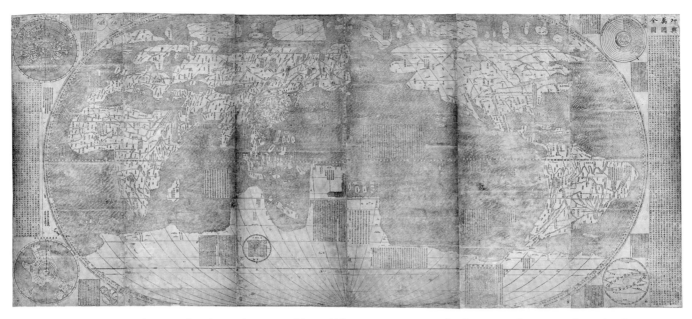

FIG. 3.2 *Complete map of ten thousand countries of the world* (Kunyu wanguo quantu), China, Ming dynasty, Wanli era, dated 1602, Matteo Ricci (1552–1610), hanging scroll (6), ink on paper, 182 x 365 cm, James Ford Bell Library, University of Minnesota.

(*tianyuan difang*).[5] Jesuit maps presented a different vision of the earth as a spherical world of many continents surrounded by seas of varying shapes. Jesuit maps also upset Chinese notions about an ethnographic space that also frames Zhuang's map. In premodern China, all peoples outside the sphere of Chinese civilization were conceived as one or another variety of *yi*, often translated as "barbarian" but better understood as "other." This model was reflected in the tributary system, a hierarchical framework of trade and diplomacy that served to define China's neighbors and to separate their cultures from Han Chinese. And keep in mind, the tribute system is mentioned specifically in Zhuang's title and thus relates directly to his map. Reiterated over centuries, these ritualized exchanges promoted the notion of a natural world "order" (*zhi*), with a Chinese imperium at the center to which satellite regimes paid tribute. The Jesuit maps, by contrast, posited a larger and unruly world, full of previously unknown others who considered themselves equal to the Chinese.[6] Yet it was precisely the Ming imperial tributary system that encouraged Chinese literati to encounter and absorb the Jesuits' alien cartographic practices. Far from a closed intellectual space, the Ming empire offered numerous points of contact where new ideas, information, methods of inquiry, and manners of thinking could gain a foothold among Chinese literati.

When Europeans arrived in China during the final decades of the Ming dynasty, their presence set off both defensive, nationalistic responses and waves of intellectual and cultural ferment throughout the region.[7] Works of faith and science, Ricci's maps were also instruments of persuasion, and Ricci deliberately cultivated friendships among Confucian scholar-officials with the ultimate aim of converting the Chinese.[8] Much Western historiography interprets the 1602 Ricci map of the world as an important milestone in a so-called "great encounter" between China and the West. Yet most Chinese intellectuals at the time treated the map as at best a curiosity and seem not to have felt the need to preserve it. It is sug-

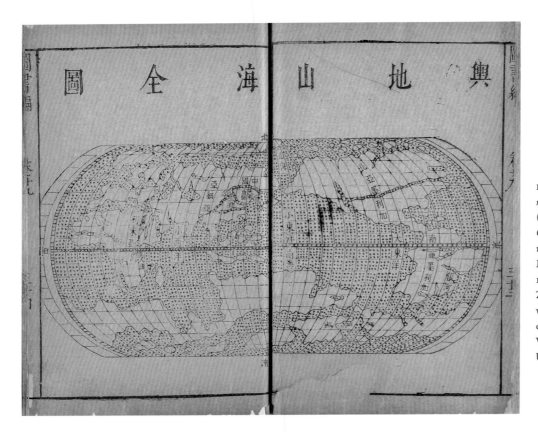

FIG. 3.3 *Complete geographical map of mountains and seas* (Yudi shanhai quantu), in *Compilation of illustrations and writings* (Tushu bian), China, Ming dynasty, Wanli era, dated 1613, Tianqi 3 (1623) reprint, Zhang Huang (1527–1608), woodblock printed book, ink on paper, 22.1 x 14.6 cm, C. V. Starr Library, Columbia University.

gestive that although Ricci claims to have printed over a thousand copies of the 1602 map, there are no extent copies in China today.⁹

Yet while Chinese of the late Ming era may not have widely valued or viewed the full-sized map of 1602, it did contribute to new notions about the shape of the earth and the configuration of its land masses. The main form in which Ricci's maps made an impact in China was through the much cruder, smaller, and more schematic variants that were frequently reprinted in encyclopedias and atlases after Ricci's death in 1610.¹⁰ Among the first was a simple outline map, published in 1613 by the Chinese scholar Zhang Huang (1527–1608) as the *Complete geographical map of mountains and streams* (Yudi shanhai quantu), in his woodblock-printed encyclopedia entitled *Compilation of illustrations and writings* (Tushu bian) (fig. 3.3).

The new Zhang map, though crude, bore the imprint of his acquaintance with Ricci in the Beijing court, that great node for cultural encounters between Jesuits and Ming literati. For the native Chinese audience, Zhang's *Yudi shanhai quantu* distilled two revolutionary concepts taken from Ricci: that the world was round, and that it was terraqueous (a series of land masses interspersed by seas). The map is thought to be modeled on Ricci's now-lost first Chinese *Map of the world* from 1584, itself based on a copy of Ortelius's *Map of the world* (Typus orbis terrarum) as featured in *Theater of the world* (Theatrum orbis terrarum), first compiled in 1570, which Ricci had brought with him to China.¹¹ Ortelius's map presents Gerard Mercator's stereographic projection as laid out in his large 1569 map of the world. Like Ricci, Zhang reversed the position of the continents to put the Pacific Ocean at the center, consigning the

Jesuit Maps in China and Korea

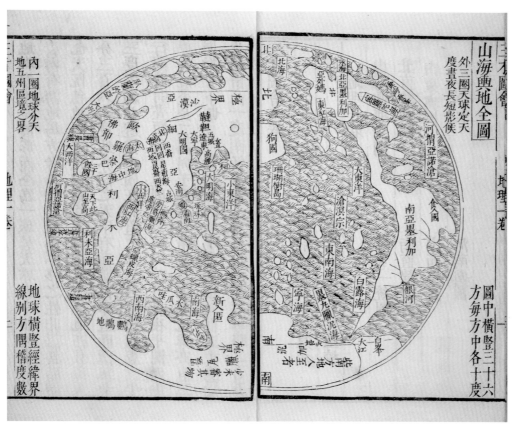

FIG. 3.4 *Complete geographic map of mountains and seas* (Shanhai yudi quantu), in *Illustrated compendium of the three powers* (Sancai tuhui), China, Ming Dynasty, Wanli era, dated 1609, Qianlong (1736–1795) reprint, compiler Wang Qi (1529–1612), woodblock printed book, ink on paper, C. V. Starr Library, Columbia University.

Americas and Europe to the peripheries at right and left. Zhang's map presents five named continents and various unnamed large islands along the bottom (south) of the map in the same oval projection Ricci introduced to China in the 1580s. In this new, Jesuit-inspired vision of the world, China remained at the center, but its context was now radically changed, as it sat among named geographies inhabited by other peoples.

Ricci's worldview also spread in the late Ming era through the popular 1609 woodblock-printed encyclopedia by Wang Qi (1529–1612), called the *Illustrated compendium of the three powers* (Sancai tuhui).[12] Figure 3.4 shows the map whose title, *Complete geographic map of mountains and seas* (Shanhai yudi quantu), is nearly identical to the title of Zhang's *Yudi shanhai quantu*. (As the map titles translate in the same way in English, they will be identified here by the names of their respective encyclopedias, *Tushu bian* and *Sancai tuhui*.)

Scholars believe that Wang Qi's *Sancai tuhui* map of the world is based on a now-lost map made by Ricci in 1600.[13] Like Zhang's map, this one preserves the same two important concepts of a round, terraqueous world. Both were typical of the simplified interpretations of the Ricci maps that most Chinese readers saw, framed as merely curious images of the world among others, since they were placed (without comment) in a sequence of world maps from different intellectual traditions.[14] Scholars have made similar arguments about the reception of the clocks brought by the Jesuits.[15] Reduced, simplified, and divorced from the European cultural context that inspired them, these world maps did not challenge the

[68] CHAPTER THREE

Chinese sense of cosmological order and otherness to the same extent that Ricci's larger format maps of 1602, 1603, and 1608 did.

This background about the Ricci-inspired Ming world maps allows us now to approach Zhuang Tingfu's 1794 map with new eyes, helping us see how this late eighteenth-century document presented a particular view of a time that now lay two hundred years in the past. The little map in the top right is none other than the *Complete geographic map of mountains and streams* (Shanhai yudi quantu), from Wang Qi's illustrated *Sancai tuhui* encyclopedia of 1609. The map in the top left is the "eastern hemisphere" map from Chen Lunjiong's (fl. 1703–1730) *Record of things heard and seen in the maritime countries* (Haiguo wenjian lu), completed in 1730 and first printed in 1744. Chen's simple schematic terraqueous map is a direct descendent, albeit only showing the eastern hemisphere, of Zhang Hong's 1613 *Complete geographical map of mountains and streams* (Yudi shanhai quantu) published in his *Tushu bian* encyclopedia.

Zhuang's two Jesuit-inspired Ming maps anchor the *gu* (or "then") of the *gujin* partnership; the larger *jin* ("now") map below was probably created as a direct result of the unsettling appearance of a British entourage in 1793. The inscription atop the lower map includes a list of tribute-bearing peoples in the period 1670–1793; the final entry in the list explains that the British arrived from the far northwest in the fifty-eighth year of the Qianlong reign (1793) and, like all tribes and states near and far, paid tribute to the Chinese emperor. This is surely referring to the well-known Macartney Mission, named from British envoy George Macartney, who was dispatched to East Asia by the British crown and the East India Company to establish the first British embassy in Beijing. Although the British embassy-building effort came to naught, someone from the ninety-plus members of the Macartney Mission likely presented maps to the Chinese. We do not know precisely which maps they were, but Aaron Arrowsmith's *Chart of the World on Mercator's Projection* (1790) or some version of his *Map of the World on a Globular Projection* (1794), both printed in London, are likely candidates. Both present the recent circumnavigations of Captain James Cook in a globular double-hemisphere projection.[16] Zhuang's map of 1794 uses the same double-hemisphere projection and includes European circumnavigation routes.

Here then was the *jin*, the troubling now from which Zhuang looked back to the first appearance of Europeans in China two centuries before. In a China now accessible to British ships that circled the globe not once but again and again, the little Ming maps had come to symbolize a complex of events, ideas, and anxieties whose full significance could only be appreciated from the later present. The Ming-era maps reminded the modern viewer of the Jesuit encounter, the West in general, and the power of Western science in particular. In league with the title and the inscription, both of which mention tribute, the new map at the bottom rolled these troubling events into the traditional package of tribute. At a moment when China's national identity in a larger geopolitics was being challenged, here was a reminder of a time in history when the tribute system—with its reassur-

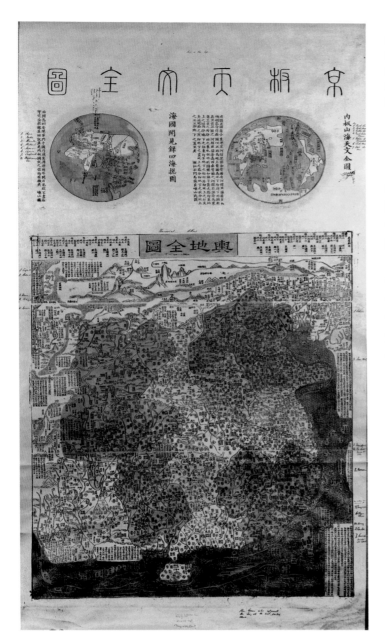

FIG. 3.5 *Capital edition of the complete map (based on) astronomy* (Jingban tianwen quantu), China, Qing dynasty, Qianlong era, ca. 1790, Ma Junliang (jinshi 1761), hanging scroll, ink and color on paper, 132 × 71 cm, Norman B. Leventhal Map and Education Center Collection, Boston Public Library.

ing "order and other"—had structured the Chinese mental world.

This moment also produced other *gujin* maps. Ma Junliang's (imperial graduate degree 1761) *Capital edition of the complete map [based on] astronomy* (Jingban tianwen quantu) also juxtaposes late Ming maps with a modern map (fig. 3.5).[17] While Zhuang's map was a one-of-a-kind manuscript, Ma's is a slightly smaller woodblock-printed edition intended for wider distribution, with numerous extant copies in the United States and elsewhere.[18]

Typically dated to around 1790, Ma's map juxtaposes the same small *Sancai tuhui* map in the top right and a small *Haiguo wenjian lu* map in the top left with the same associated text as found on Zhuang's map.[19] The large map below is one from the Huang family lineage of Qing court mapmakers.[20] It derives from Huang Zongxi's (1610–1695) 1673 *Map of China* (Yudi quantu), amended by his grandson Huang Qianren (1694–1771) in 1767. The date is important, for just a few years before, in 1759, Manchuria, Mongolia, Tibet, and Chinese Turkistan were finally brought under Qing rule. Emperor Qianlong appointed officials from the Manchu administrative units called banners, as well as local political figures, to supply statistics about their administrative districts.[21] That information was compiled by Huang Qianren and published as a multicolored woodblock-printed map in 1767 with the title *All under heaven map of the everlasting unified Qing empire* (Daqing wannian yitong tianxia quantu). Ma presents his version of the Huang family map on the *Jingban tianwen quantu*.

Although we cannot know for sure, Ma's map also likely dates to the middle of the 1790s. Like Zhuang's, it connected "then and now" visually, looking to the past to find accessible memories that matched and helped explain the circumstances of the present. Zhuang made his purpose obvious by using the term *gujin* in the title, by referring to the tribute system,

[70]   CHAPTER THREE

and by mentioning Ricci in his inscription. Zhuang's map was a manuscript produced for imperial consumption, while Ma's map was a woodblock print marketed to a wider audience. Regardless, they were produced at essentially the same time and both reflect the same moment in Chinese history: one when the Jesuit-inspired notions of world geography introduced during the late Ming were being revisited and set side-by-side with (if not fully integrated into) indigenous Chinese systems of geographical thought as the Chinese came to terms with a largely unknown and potentially threatening world once again in the 1790s.

Jesuit Maps in Korea

Although the Jesuits did not directly visit Korea, their cartographic legacies also unfolded there in ways that show distinct variations of the *gujin* paradigm. Premodern Koreans inherited a cartographic history that was heavily Sino-centric. This is exemplified in the best-known early Korean map of the world: the magnificent hand-painted wall map from 1402 entitled *Map of the integrated regions and capitals of states over time* (Honil gangni yeokdae gukdo jido, often abbreviated Gangnido).[22] The *Gangnido* presented a single landmass encircled by water in a flat two-dimensional plane, in keeping with the Chinese "round heaven, square earth" cosmology.

Jesuit cartographic history came to Korea indirectly, since the ruling Yi family of Korea's Joseon dynasty (1396–1910) forbade Jesuits from entering Korea. Nonetheless, the Korean court kept abreast of Jesuit science, acquiring copies of Ricci's maps and related documents from the early seventeenth century. According to the *Annals of the Joseon dynasty* (Joseon wangyo sillok), for instance, the Korean envoy Yi Gwangjeong (1552–1629) acquired copies of Ricci's 1602 and 1603 maps at the late Ming court in Beijing and brought them back to Seoul along with other books.[23] Thus we know that Jesuit maps were seen in the Korean court not long after they appeared in China, although there is less evidence of how Korean officials regarded them.[24]

A century later, in 1708, Jesuit maps of the late Ming were copied in Korea by royal order. At the command of King Sukjong (r. 1674–1720), the Royal Astronomy Bureau made copies of Ricci's 1608 *Kunyu wanguo quantu* (K. Gonyeo man'guk jeondo)[25] and Adam Schall von Bell's 1634 star chart entitled *Two general maps of the stars relative to the ecliptic* (Huangdao zongxingtu).[26] There are probably two reasons why the Joseon court wanted reproductions of late Ming Jesuit maps.[27] First, according to the official court chronicle, *The daily records of the royal secretariat* (Seungjeongwon ilgi), in the fifth month of 1708 the court wanted to correct discrepancies in the official Joseon calendar. This entailed consulting Chinese calendars and star charts that had been produced in the late Ming and early Qing under Jesuit supervision.[28] Second, Korea was undergoing a surge of Ming loyalism and nostalgia at the time—encouraged by the government's ideological program to establish the Joseon dynasty as the legitimate heir of the fallen Ming (at a time when China was ruled by a non-Han conquest dynasty, the Man-

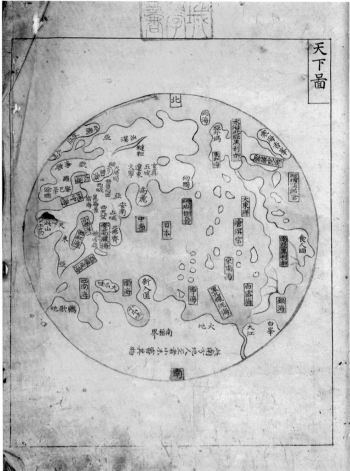

FIG. 3.6 *Map of the world* (Cheonhado), in *Atlas* (Yeojido), Korea, Joseon dynasty, late eighteenth century, woodblock print, ink and color on paper, 36.8 x 28.6 cm, Kyujanggak Archive, Seoul National University.

chu). These maps were produced to coincide with the sixtieth anniversary of the Ming dynasty's fall, and they anchor a *gujin* variant that reaches out to a Chinese Ming historical moment to confirm a current position in the Korean court. With their origins in late Ming China, these Jesuit maps served as a reminder of the ideal Chinese principles of "order and other" as articulated by the Ming dynasty, connecting an idealized Chinese past to the ruling regime in Korea.

Interest in late Ming Jesuit maps subsided for the next eighty years. When it resurfaced in the late eighteenth century, it did so in the context of manuscript and printed atlases, or *Yeojido*, in a second variation of the *gujin* paradigm. The agenda of these Korean atlases, which typically contained thirteen maps, was to confirm Korean identity by portraying Korean space in a number of different cartographic frameworks at different scales.[29] Within these *Yeojido*, two types of world maps (known as *Cheonhado*) emerged. The first was a highly schematic "circular" or "wheel" type; the second, the so-called "world" type, developed during the late eighteenth century and reemerged in the late nineteenth century.[30] The "world" *cheonhado* (fig. 3.6) is a copy of the late Ming *Sancai tuhui* map but here with the title *Map of the world* (lit. "map of all under heaven") (Cheonhado).

The atlas format was created for the Korean court during the 1790s, at the same time as the *Sancai tuhui* maps reappeared in the Chinese court. Its appearance in Korea is likely linked to its appearance in the Chinese court, suggesting a *gujin* variation borrowed directly from China. This specific *Sancai tuhui*–style map, found only in Korean atlases in the late eighteenth century, was produced in very limited quantity (fig. 3.7). Atlases continued to be made, but the map of the world included in later atlases shifted to a new, unique "wheel" type of *cheonhado*. Strikingly, however, the older "world" *cheonhado*, based on the *Sancai tuhui* map, reappeared a hundred years later—precisely when Western powers established their first direct diplomatic ties in Korea proper.[31] For example, the atlas that this "world" *cheonhado* comes from was acquired by George Clayton Foulk while acting as the first US naval attaché to Korea in the 1880s.[32]

The Chinese and Korean maps from the late eighteenth century show how literati of that era invoked

the past to cope with anxieties about the present. With the West intruding into East Asia, Chinese and Korean mapmakers gazed backward to an earlier moment in Chinese history, when late Ming scholars and courtiers had confronted an earlier group of visitors, the Jesuits, who had brought an exotic array of maps that were slowly but incompletely assimilated into Chinese and Korean cartographic practice. In both eighteenth-century China and Korea, Jesuit cartography served to connect the first and second coming of the West to the region.

Thus, to fully appreciate China and Korea's engagement with Jesuit cartography, we must consider them within the context of Chinese notions of time and space. Only then can we appreciate the distinctive temporal message that late Ming Jesuit maps conveyed when they were resurrected, copied, and reprinted in China and Korea nearly two hundred years after their creation. The Jesuits were long gone, but their historical moment echoed within the maps, an enduring symbol of transcultural contact and exchange. The *gujin* paradigm points to the active deployment of remembered events, with all the emotional and intellectual associations that they carry, to connect to and explain a present moment. The turn to Jesuit-inspired maps in China and Korea during the eighteenth and nineteenth centuries suggests that they had come to constitute a visual paradigm for contact with the West. These *gujin* maps encouraged the literate public to grapple with clashing notions of "other and order" and to remap their place in the larger world. The late Ming Jesuit maps became an emblem of the West's return, foretelling an end

FIG. 3.7 *Map of the world* (Cheonhado), in *Atlas* (Yeojido), Korea, Joseon dynasty, late nineteenth century, woodblock print, ink and color on paper, H. 23 cm, American Geographical Society Library (469 B), University of Wisconsin, Milwaukee.

of isolationism, and a reconsideration of time, space, identity, and place in the world.

This temporal messaging exemplifies the aims of this volume, showing how often maps have presented a dynamic world in a state of flux, rather than static images of what William Rankin (in this volume) calls "an eternal present." Chinese and Korean maps in the *gujin* tradition show how the late Ming past was not dead in the eighteenth century, but very much alive

Jesuit Maps in China and Korea   [73]

in Jesuit-inspired maps that helped thinkers of the time decipher patterns and paradigms in the present. These Ming map revivals themselves reveal the particular qualities of the Chinese intellectual tradition, in which Jesuit maps served to "re"-present the past in a Chinese way, serving as a spatiotemporal location-device with which to recognize and respond (once again) to the challenges of a globalizing world.

Notes

1. Richard A. Pegg, *Cartographic Traditions in East Asian Maps* (Honolulu: MacLean Collection and University of Hawai'i Press, 2014), 35–42. The final five characters of the title of the Library of Congress map are missing, due to paper loss. The 1800 copy of this map and later woodblock-printed copies, both now in the Maclean Collection, provide the final missing characters.

2. Stephen Kern, *The Culture of Time and Space, 1880–1918* (1983; repr. Cambridge, MA: Harvard University Press, 2003); and Kuang-ming Wu, "Spatiotemporal Interpretation in Chinese Thinking," in *Time and Space in Chinese Culture*, ed. Chun-chieh Huang and Erik Zürcher (Brill: Leiden, 1995), 18–19.

3. Huang Shijian and Gong Yingyan, *Li Madou shijie ditu yanjiu* (A Study of Matteo Ricci's World Maps) (Shanghai: Shanghai guji chubanshe, 2004), 3–41.

4. For a bibliography of the scholarship on Ricci maps, see Qiong Zhang, *Making the New World Their Own: Encounters with Jesuit Science in the Age of Discovery* (Leiden: Brill, 2015), 29–30n7.

5. Cordell D. K. Yee, "Taking the World's Measure: Chinese Maps between Observation and Text," in J. B. Harley and David Woodward, eds., *The History of Cartography*, vol. 2, bk. 2: *Cartography in the Traditional East and Southeast Asian Societies* (Chicago: University of Chicago Press, 1994), 96–127.

6. Mark Lewis, *The Construction of Space in Early China* (Albany, NY: SUNY Press, 2005).

7. David E. Mungello, *The Great Encounter of China and the West, 1500–1800* (1999; repr. Rowman & Littlefield, 2013).

8. Liu Yu, *Harmonious Disagreement: Matteo Ricci and His Closest Chinese Friends* (New York; Bern: Peter Lang, 2015), 193.

9. The six known copies that survive are housed respectively in the Vatican, the University of Kyoto, the Miyagi Prefecture Library, the Diet Library in Tokyo, a private collection in Paris, and the James Ford Bell Library at the University of Minnesota.

10. These simplified versions were possibly closer to Ricci's own (lost) earlier and simpler maps of the world, reportedly made in 1584 and 1600. See Huang, *Li Madou shijie ditu yanjiu*, 3–41. Yee challenges the notion that Western mapmakers had much influence on Chinese mapmaking from the late Ming through the Qianlong reign of the Qing empire. For Yee, Western mapmaking did not truly take hold until after 1842, when the Chinese Navy was defeated by the British during the first Opium War. See Yee, "Traditional Chinese Cartography and the Myth of Westernization," in Harley and Woodward, ed., *History of Cartography*, vol. 2, bk. 2, 191.

11. Richard Smith, *Chinese Maps* (Oxford: Oxford University Press, 1996), 43–44; and Cao Wanru et al., "*Zhongguo xiancun Li Madou shijie ditu de yanjiu* (A Study of the Extent Copies of Ricci's World Maps in China)," *Wenwu* 12 (1982): 57–59.

12. This map has the same characters in the title but in a different order from the map associated with Ricci's original titles for his 1584 and 1600 maps of the world. The difference is negligible in Chinese and English. See Yee, "Traditional Chinese Cartography," 171.

13. Yee, "Traditional Chinese Cartography," 175.

14. Matthew W. Mosca, *From Frontier Policy to Foreign Policy: The Question of India and the Transformation of Geopolitics in Qing China* (Stanford, CA: Stanford University Press, 2013).

15. Catherine Jami, "Western Devices for Measuring Time and Space," in *Time and Space in Chinese Culture*, 169–200.

16. Richard Pegg, "World Views: Late 18th Century Approaches to Mapmaking in China and Britain," *Orientations* 44, no. 3 (April 2013): 84–89. Another work depicting circumnavigations is Robert Sayer, *A General Map of the World* (London, 1787).

17. Ronald E. Grimm and Roni Pick, *Journeys of the Imagination: An Exhibition of World Maps and Atlases from the Collections of the Norman B. Leventhal Map Center at the Boston Public Library, April 2006 through August 2006* (Boston: Boston Public Library, 2006), plate 36. Ma's maps measure 132 x 71 cm.

18. Copies can be found in Rice University, Fondrian Library (G7821.A5 1761 M3), Library of Congress (G7820 1790.M3), and a private New York collection that came out of Sotheby's London, Auction L16409, lot 213, Nov. 7, 2016. The provinces of China for these three versions are not colored. Some versions do not have the *yudi quantu* title on the lower map, the abbreviated title of those found on the *sancai tuhui* and *tushu bian* maps.

19. Ma Junliang along with Lin Binglu later reprinted Chen Lunjiong's *Haiguo wenjian lu* (Record of Things Heard and Seen in the Maritime Countries) in 1793.

20. Bao Guochiang, "Qing Qianlong 'Daqing wannian yitong tianxia quantu' bianxi" (Study of Qing dynasty Qianlong's *All under Heaven Map of the Everlasting Unified Qing Empire*), *wenxian yanjiu* (Archival Studies), no. 2 (2008): 40–44.

21. In addition, since the late seventeenth century the Jesuits had been involved in various court-sponsored map surveys of these areas. See Mario Cams, *Companions in Geography: East-West Collaboration in the Mapping of Qing China (ca. 1685–1735)* (Leiden: Brill, 2017).

22. The original is lost. The Tenri Central Library in Japan has a copy that has been dated to ca. 1568. See Gari Ledyard, "Cartography of Korea," in *History of Cartography*, vol. 2, bk. 2, 244–49, fig. 10.3. For an accessible account of this remarkable map, see Jerry Brotton, *A History of the World in 12 Maps* (New York: Viking, 2013), chapter 4; on Korean cartography more generally, see Yong-U Huan et al., *The Artistry of Early Korean Cartography* (Larkspur, CA: Tamal Vista Publications, 1999).

23. M. Antoni J. Ucerler, "Missionaries, Mandarins, and Maps: Reimagining the Known World," in *China at the Center: Ricci and Verbiest World Maps*, ed. Natasha Reichle (San Francisco: Asian Art Museum, 2016), 1–16.

24. Copies of the *Sancai tuhui* were also acquired at this time. Yi Sugwang (1563–1628), the assistant superintendent in training in the Institute for the Advancement of Literature, traveled to Beijing several times starting in the 1590s. According to his encyclopedia entitled *Jihong yuseol* (Topical Discourses of Jihong), during his 1614 trip to Beijing he acquired numerous books, including a copy of the *Sancai tuhui*. Huang, *Li Madou shijie ditu yanjiu*, 118. Yi considered the Ricci maps in the category of exotica, placing his remarks in a chapter devoted to marvels and strange things from remote foreign lands.

25. Today, the Kyujanggak Archive of Seoul National University, formerly the Royal Library of the Joseon dynasty (1392–1910), preserves two 1708 copies of Ricci's maps commissioned by the Joseon government. One is dated the eighth month, and the second, made in the ninth month of 1708, is preserved in photographs taken in 1932 from a now-lost copy. It is generally believed that the original map used to make the copy was one of the original twelve presented to the Ming Wanli emperor in 1608. See O Sanghak, *Joseon sidae segye jido wa segye insik* [World maps in the Joseon dynasty and the perception of the world/The Joseon dynasty's world maps and its perception of the world/Its worldviews] (Seoul: Changjak gwa bipyeongsa, 2011), 181.

26. F. Richard Stephenson, "Chinese and Korean Star Maps and Catalogs," in *History of Cartography*, vol. 2, bk. 2, chap. 13, 570–72, fig. 13.39.

27. See Lim Jongtae, "Matteo Ricci's World Maps in Late Joseon Dynasty," *Korean Journal for the History of Science* 33, no. 2 (2011): 277–96; and Lim Jongtae, "Learning Western Astronomy from China: Another Look at the Introduction of the *Shixian li* Calendrical System into Late Joseon Korea," *Korean Journal for the History of Science* 34, no. 2 (2012): 197–217.

28. In China, the Chongzhen calendar reforms in the 1630s using Jesuit science had made corrections in the late Ming and again early in the Qing dynasties. These reforms were not entirely embraced by the Joseon court at that time, as the Korean court did not formally recognize Qing rule and therefore did not incorporate all the updates. By 1705, however, calculations to planetary movements needed corrections. Chief state councilor Choe Seokjeong (1645–1715) supervised the 1708 reproduction of Ricci's world map and Adam Schall von

Bell's star map. Choe placed the two Jesuit works into the ideological framework of contemporary Ming loyalism. See Kim Seulgi, "Sukjong-dae Gwansanggam ui siheonnyeok hakseup: Eulyu'nyeon yeokseo sageon gwa Geu e daehan Gwansaggam ui daeeung eul jungsim euro," master's thesis, Seoul National University, 2016.

29. First within the world as seen in a *Map of the World* (commonly referred to as a *Cheonhado*), then within East Asia with a *Map of China*, a *Map of Japan*, or a *Map of the Ryukyu Islands*, then locally with an overall *Map of Korea*, and finally as the sum of its constituent parts through a map of each of the eight provinces (*paldo*) with administrative bodies and general typography and finally the capital region and the capital city itself. Richard A. Pegg, "Maps of the World (*Cheonhado*) in Korean Atlases (*Yeojido*) of the Late Joseon Period," in *Arts of Korea: Histories, Challenges and Perspectives* (Gainesville: University of Florida Press, 2018), 286–303.

30. Pegg discusses these "wheel" and "world" types of *Cheonhado* in detail ("Maps of the World"). Korean cartographer Yi Chan coined the term "wheel" *Cheonhado*. In an alliterative response in English, I have coined the term "world" *Cheonhado*.

31. The Korean court was aware that in the 1860s British and French troops had occupied Beijing, and that soon after China and Japan were both forced to accede to Western demands on trade and diplomacy. Korea was no longer isolated, and by the 1880s the West had established diplomatic relations.

32. Given the atlas's pristine condition, it is fair to assume it was printed at the time that Foulk was stationed in Korea. See Samuel Hawley, *Inside the Hermit Kingdom: The 1884 Korea Travel Diary of George Clayton Foulk* (Lanham, MD: Lexington Books, 2008). Other equally pristine Korean atlases also acquired by Westerners in the 1880s can be found in the British Library (Or15965) and Hamburg Ethnographic Museum (82.93.1). See Richard Pegg, "Maps of the World," 288.

PART II

The Atlantic World

The name of this section—"The Atlantic World"—is itself a space-time concept. It gained traction in the post–World War II era, when NATO nurtured a larger Atlantic vision among war-weary nations. Since then, academic historians have used the term to refer to the three centuries from roughly 1492 to 1800, when Europe, Africa, and the Americas were knit into a larger—albeit fractious and fractured—unit. Suddenly new worlds opened for millions of people around the Atlantic Ocean. And whether facing that ocean from east or west, they encountered one another through a process of creativity, destruction, and synthesis.

Radically new concepts of time and space emerged from this post-Columbian encounter, as people around the Atlantic tried to make sense of the old and the new. This section shows how maps were some of the messengers for these novel temporalities. The atlas, to take an obvious example, invites the reader to experience in real time a vast chronological canvas by paging through a succession of historical and contemporary maps. And beyond the realm of the strictly cartographic, other graphic systems such as the timeline emerged to represent time's passing in new ways.

This section opens with Barbara Mundy's exploration of Aztec maps created soon after the Spanish conquest of 1519–1521. In addition to explaining the fascinating time-space configurations of these maps on their own terms, Mundy also shows how the Aztec maps offered implicit critiques of the visions of time and space being imposed by the Spanish colonial regime. She reveals that centuries-old Aztec maps resurfaced again during the moment of Mexican independence in the nineteenth century, providing a new "regime of historicity" for a young nation.

Veronica Della Dora's essay moves us across the Atlantic to reveal that European mapmakers in the seventeenth and eighteenth centuries also wrestled with imperialism's impact on traditional notions of time. New discoveries pouring into Europe from all over the world dissolved the closed and divinely ordained Aristotelian cosmos dominant since classical antiquity. In its place, the Scientific Revolution installed an infinite universe bound by fundamental laws—including Newtonian notions of time as linear and universal. Della Dora shows how mapmakers, sculptors, and painters deployed a wondrous variety of veils and curtains to make palpable Time's role in revealing and concealing new knowledge.

Taking off from Michel Foucault's insight that languages are conceptual maps, Daniel Rosenberg's essay explores an extreme case of the spatiotemporal innovation in early modern Europe: maps of language. Polymaths such as John Wilkins—a founder of the Royal Society—cherished the utopian hope that a simple, logical, artificial language could replace the confusing irregularities of natural languages, which were now being revealed in their hundreds of varieties by the age of exploration. Rosenberg highlights the maplike quality of Wilkins's diagram of prepositions, which ingeniously attempted to locate words such as "after" and "before" in relational space.

The three essays here—each in their own way—remind us that both time and space are social constructions, powerful cognitive tools that allow groups of people to orient themselves in relationship to one another, to nature, and to the metaphysical world. The essays help explain why maps of time changed so quickly in the period 1492–1800, an age of massive disorientation for so many.

4

Barbara E. Mundy

History in Maps from the Aztec Empire

Maps and calendars are the Janus faces of a single phenomenon: the need of human societies to organize their collective existence by making visible, and bringing into the horizon of cognition, the otherwise immaterial ideas of space and time. "Among the symbols which human beings can learn, and from a certain stage of social development on, must learn as a means of orientation," Norbert Elias has pointed out, "is time."[1] He goes on to emphasize that time is a social construct: that is, "a frame of reference used by people of a particular group . . . to set up milestones recognized by the group within a continuous sequence of change, or to compare one phase in such a sequence with phases of another."[2] He posits that the "social regulation of time" by means of calendars and clocks (be they of the grandfather or life variety) is a fundamental mechanism through which societies cohere, synchronizing activities of the group (the time for planting, for instance) and providing constructs to understand their lived experience within time, perceived as a flow. If a calendar is a mode through which time is structured and expressed, the same can be said of a map in relation to space. Space, as Edward Casey has argued, can only be experienced by human beings as place—individual and specific, rather than abstract and general.[3] Like others, Casey has posited that our perceptions of space are determined by the social context in which they occur; space cannot be known except as place.[4] By these lights, the map, by ordering space along understood and socially determined coordinates, is a structural parallel of the calendar. Both these authors, the sociologist Elias and the philosopher Casey, join in rejecting a Cartesian assumption that time and space are a priori phenomena, existing before and beyond human attempts to describe and analyze them. For both thinkers, space and time are social constructs: powerful tools that allow people to orient themselves vis-à-vis one another and the non-human world, whether it be called "nature" or "landscape" or "outer space."

It thus comes as little surprise that maps and calendars are to be found across the globe. Both have structural similarities: they rely on measuring practices, and depend on graphic abstractions for expression. One feature of modernity in the West has been to set each form on a trajectory toward greater and greater precision in measurement, often involving elaborate technological means. For instance, early modern Europe had inherited the Julian calendar

from the Romans, which included a leap day every four years. In 1582, after mathematicians and astronomers had revealed that calendar's imprecision, Pope Gregory XIII promulgated a new calendar, omitting three leap days every 400 years to bring the calendar into greater alignment with the length of the earth's revolution around the sun. In the same century, cosmographers and mapmakers worked to correct for magnetic declination in cardinal orientations, and worked out the trigonometry to measure long distances on land; today, distances are measured by laser, not rods. The rhetoric around this quest blinds us to the fact that both calendar and map, as graphic expressions, are tracking socially determined referents. While the earth's rotation around the sun is not a social construct, our choice to use the measurement of that cycle as the foremost organizing principle *is*. Likewise, what we perceive as "mappable" space is also the product of social decision-making. If we glance at a conventional world map, the oceans, those watery worlds, appear as vast expanses of largely uniform blue. Despite modernity's quest for the accurate map and the perfect calendar, maps and calendars share a socially determined referent—and therefore, a perennially unstable one. And as the chapters in this volume show, maps and calendars are frequently yoked together in societies across the world.

That was certainly the case in indigenous America. This chapter considers a series of maps created in the sixteenth century by a people popularly known as the Aztecs, who had built an extraordinary empire over the course of the fourteenth and fifteenth centuries in central Mexico. This autochthonous empire was brought to an end by the Spanish conquest of 1519–1521. These graphic works were all made after the conquest, and on one hand, they provide evidence of the generality discussed above—they are all concerned with both time and space. On the other hand, they offer us a distinct perspective on how at least one group of indigenous Amerindians chose to graphically structure time and space, thereby revealing the socially constructed nature of each. Created in a limited time period, they are also marked by the particularities and pressures of a distinct historical context. When these maps were produced, Spanish conquistadors were expropriating valuable lands, and evangelizing Catholic friars were imposing new ritual calendars; as a result, Aztec social constructs of both space and time were being radically transformed in the colonial regime. In preserving an earlier, alternative vision of both space and time, the indigenous mapmakers of the sixteenth century offered an implicit critique of the ways that time and space were changing around them. Their critique proved consequential in ways they never anticipated: by the 1850s, following Mexico's war of independence from Spain, Mexican intellectuals reencountered indigenous maps in antiquarian collections. They used them, along with Aztec artworks, to establish a "regime of historicity" for their new nation; that is, to articulate newly minted Mexico's past, present, and future, and thereby to authorize and justify its existence and planned trajectory.[5]

Creation Stories and the Linkage of Time and Space

Extant Aztec sources, most recorded during the sixteenth century, pay little attention to the deep origins of time and space. Rather, they are concerned with the creation of the kind of time and space that allows humans to exist and thrive. An important collection of Aztec creation histories, known in a French translation, the "Hystoyre du Mechique," establishes the following.[6] After four sequential cycles of creation and destructions of worlds, each with its own sun, each unfit for human existence, two male creator deities, Tezcatlipoca and Ehecatl, decided to create a new world.[7] They identified a "goddess" (*déesse*), named Tlaltecuhtli or "Earth Lord," and, turning themselves into serpents, ripped her body in two.[8] One half of the dismembered deity formed the earth, the other half the sky. Upon seeing the violence done to Tlaltecuhtli, the other deities were aggrieved and wanted to repay the earth deity for the damage done to her. So they transformed the spaces of Tlaltecuhtli's body—which offered orifices in the form of caves, and protuberances in the form of hills—by adding the plants, trees, and water necessary for human survival. With this, habitable space came into being.

The Aztec cosmogony sets the creation of space and the creation of time in the same sequence of events, and in doing so, reveals something of the indigenous understanding of each. Space was habitable space, and time was movement. The "Hystoyre" continues to recount that, even after the creation of habitable space, the earth was static and in darkness because there was no heat- and light-giving entity to illuminate the days and to create time.[9] So the deities joined together to offer riches to create the sun. One of the deities, the poor, pockmarked Nanahuatzin, had no great gifts to offer, so he hurled himself into a bonfire in self-sacrifice. His fiery body rose to become the sun, but one element was missing: movement.[10] The sun stood still in the sky, doing more harm than good by scorching the earth. It was only through further sacrifices on the part of the deities that the wind—an invisible moving force—pushed the sun into motion.

As this brief incursion into cosmogony suggests, the creation of space and the creation of time were linked in the understanding of indigenous people of Central Mexico, as they were in other places and times in the world as well. (As in other theocentric cosmologies, the origin of the divine is taken as given.) More specific to the region was the emphasis on *habitable* space and the understanding of movement and change *as* time, as well as their fundamental linkage: it is only with the creation of a space fit for humans that movement and change—in other words, time—could begin.[11] Also distinct in the Americas is the idea that the proper orientation of human action was to mirror the actions of the deities, so that histories like the "Hystoyre" were not only explanations of origins but also guides to practice. If the deities made sacrifices of themselves to create the sun, humans needed to make sacrifices as well.[12] If Ehecatl and

Tezcatlipoca's violent treatment of a goddess's body brought the world into being, humans too could resort to violence for creative ends.

In short, a cosmogony like the "Hystoyre" registered the actions of deities in world-making. Cartography and calendrics were related to those divine acts in that they registered similar creative actions on the part of humans. Maps showed an arrangement of features deriving from both generative moments, emphasizing the caves and hills that were part of the earth deity's body but also including named places that human beings had made. And calendars showed the arrangement of days and years that divided the flow of time into the phased sequence that, in Norbert Elias's view, *is* time.

The "Hystoyre du Mexique" reiterates the ontological sameness of space and time by including a calendar, the only graphic rendering on its pages (fig. 4.1). Similar renderings, like that found in the Tovar Calendar of ca. 1585 (fig. 4.2), appear in a number of sixteenth-century manuscripts produced in Central Mexico, so the graphic rendering of "Hystoyre du Mexique" wherein time and space are connected is probably not a novelty, but a convention with deep history.[13] Both express the interconnectedness of space and time (or maps and calendars) in ancient Mexico. Figure 4.2 presents a calendar of fifty-two years. Every solar year was given one of four possible names—Rabbit, Reed, Flint Knife, House—and these are represented by conventional glyphs, with rabbits in the blue sections, reeds in the green sections, flint knives in the red sections, and houses in the yellow sections. The count begins at the sun at the center (the origin of space and time), and above it appears the first year, 1 Reed (the ordinal is represented by a single white circle). Counterclockwise, moving to the red arm, one finds the second year, 2 Flint Knife, and then, on the yellow arm, the third year, 3 House. The years followed in a set succession with the ordinal changing with each new year, to continue with 4 Rabbit, 5 Reed, 6 Flint Knife, and so on. After 13 Reed, the ordinal count reverts to 1, which is now paired with Flint Knife to name the year 1 Flint Knife, thus creating a new sequence of number-name combinations. By continuing this process four times, the ordinal 1 will be once again paired with Reed after fifty-two years (13 ordinals x 4 year names), and the cycle begins again. (Another calendar, the ritual calendar of 260 days, is not explicitly included in the diagram. It was likely derived from the length of time of human gestation, from the time of missed menses to birth.)[14]

The "Hystoyre du Mexique" presents a similar diagram of the cycle of the years, but it is radically simplified to show just four years, each set at the end of the arms of a cross, also following the same counterclockwise order as in the Tovar Calendar. What makes this diagram notable is that each of the four years is connected to one of the cardinal directions. From the center right, moving counterclockwise, it reads (in both Nahuatl, the Aztec language, and French, which I have set in italics): "1 Rabbit / *South* / South; 2 Reed / *East* / East; 3 Flint Knife / *North* / North; 4 House / *West* / West."[15] Rabbit years were connected with the south (called *Huitztlan* in the Nahuatl gloss), Reed years with east (*Tlapco[pa]*),

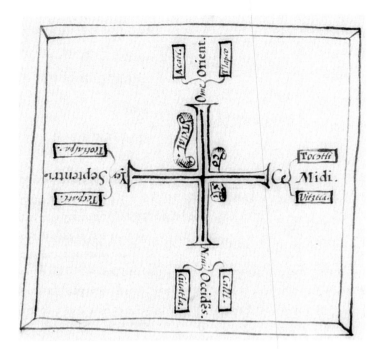

FIG. 4.1 Count of years, "Hystoyre du Mechique," sixteenth century. Ink on European paper, 29 × 20.2 cm. "Fragments d'André Thevet sur les Indes occidentales et sur le Mexique." Français 19031, fol. 82v, detail, Bibliothèque Nationale de France. Artwork in the public domain, photograph courtesy of the Bibliothèque Nationale de France.

FIG. 4.2 Unknown creator, Calendar of 52 years, Tovar Calendar, ca. 1585. Pigment on European paper, 21 x 15.2 cm. Juan de Tovar, "Historia de la benida de los yndios apoblar a Mexico de las partes remotas de Occidente." Codex Ind 2, fol. 142r, John Carter Brown Library. Providence. Artwork in the public domain, photograph courtesy of the John Carter Brown Library.

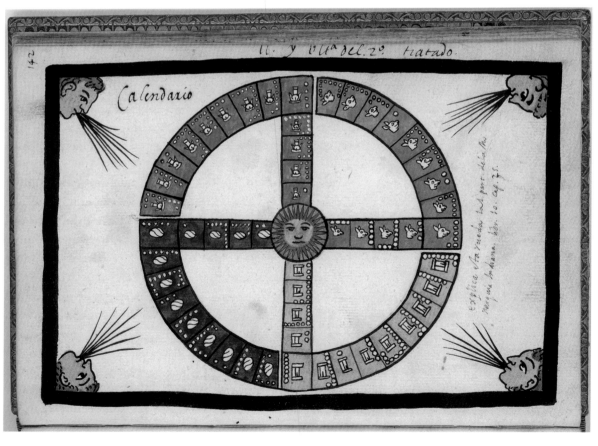

Flint Knife years with north (*Teotlalpan*), and House years with west (*Cihuatlan*). The etymology of the Nahuatl shows that at least three of the "directions" were actually conceived of as places. "Huitztlan" means, roughly, "Next to the thorns." Cihuatlan means "Next to the woman," called such because women who died in childbirth were believed to rest in the west, where the sun sets. "Teotlalpan" means "At sacred lands."

The simplicity of the graphic of the fifty-two-year solar calendar reveals little of its wide spread and deep origins, but archaeologists and ethnohistorians have established that the fifty-two-year solar calendar, along with these directional associations, was found throughout Mesoamerica. This culture region roughly corresponds to Central Mexico and Guatemala and contained the great urban cultures of the North American continent, the earliest of its cities having been founded in the first millennium BCE. The calendar was also of great antiquity, with calendrical dates found on stone monuments from the first millennium BCE.

The conceptual unity of space and time meant that, in the larger Mesoamerican tradition, cartography was often linked to the calendar, establishing one frame of reference for the movement of time. But even more frequently it was linked to historical narrative, which shows the movement of time through the meter of human action.[16] In the examples that follow, I will explore this linkage between maps and historical narrative to emphasize the conceptual union of space and time. One strand of argument is that time *was* movement, so the historical events appearing on the surfaces of maps express their diachronic nature through symbols for movement, like pathways and footprints. As suggested in the creation history of the "Hystoyre du Mexique," when mapmakers set down place and time on the flat surface of the manuscript page, they were mirroring the actions of the deities at creation, and were, no less than the deities, involved in world-making. Two kinds of maps from Central Mexico, in particular, offer two facets of the process of world-making: the map of migration, and the map of conquest.

World-Making through Migration

Most peoples of Mesoamerica believed themselves, or their rulers, to have had special beginnings. The group we know as the Aztecs, who would eventually found Tenochtitlan (the city that lies under today's Mexico City), held that they had originated on an island quite distant from Tenochtitlan called Aztlan, this point of origin giving rise to their name. While ethnic groups like the Aztec may have originated elsewhere, the spaces that they would come to occupy were seen to have been preordained. Given that originating sites lay far away from current settlements, one function of cartography was to document the connection between the two, to show how a people had traveled, through space and over time, to arrive at what was destined to become home. This transition was a crucial one across Mesoamerica, and some maps show early ethnic leaders dressed in rude animal skins, whereas their descendants, as settled peoples, would adopt "civilized" garments of woven

fibers. And settlement meant creating an *altepetl*, a Nahuatl term that translates into "water hill" but which means more broadly an autonomously ruled territory and people. At the time of the Conquest, the great *altepeme* (the plural of *altepetl*) of Central Mexico comprised large cities, which had as many as 50,000 to 100,000 inhabitants.

The so-called Mapa de Sigüenza, named for an early owner, is a map that documents the migration out of Aztlan by the Aztecs (fig. 4.3).[17] On a single sheet of indigenous *amatl* paper, it maps this thereto-here journey, a journey that took place over 150 years.[18] The narrative largely follows the path of the footprints that move between the "there" and the "here," the "there" being Aztlan, whose place name dominates the upper right quadrant, and the "here" being the Valley of Mexico, which fills the lower left quadrant. Aztlan, seen in figure 4.4, is rendered as a square lake, filled in by wavy blue lines to represent currents. In the center of this lake, a blue hill rises, and from it a five-limbed tree grows, its needlelike foliage (but not its trunk structure) reminiscent of a native cypress. At the top of the tree, an eagle spreads its wings. A cloud of hook shapes appears before its beak to indicate that it is speaking. Gathered to listen are a group of ten native elders, leaders of the Aztecs.

In its moment, the pictorial narrative would have been fleshed out by an oral account; in our moment, we depend upon alphabetic texts, written down after the Conquest, to guide our understanding (the pitfalls of our reliance on text would make their own essay). These texts tell how the tribal deity, Huitzilopochtli, in the form of a bird, called on the leaders of the Aztecs to leave Aztlan to found a new settlement. That journey would take the Aztecs through different regions (probably ones to the north of the Valley of Mexico) before they arrived as unwelcome newcomers into the well-populated Valley.[19] Their history is told by compact images of two sorts: a hieroglyphic place name accompanied by blue discs, symbols for solar years, to show the place and the amount of time spent there; and schematic representation of a notable event. The sequence of the history is signaled by footprints to show the movement in time and across space of the principal actors. For instance, figure 4.5 shows us a place—Chapultepec. In Nahuatl, the place name means "Hill of the Grasshopper," and here it is rendered as a curved hill upon which a grasshopper perches. Three roads, all marked with footprints, converge on the hill to show the paths of narrative action, rather than any set path in the landscape; the four blue discs at the left show that the Aztecs spent four years at Chapultepec. They were soon chased out by the Tepanecs, who controlled much of the Valley, and in the battle with the Tepanecs, Aztec warriors were killed. On the map, this part of the narrative is compressed into a single moment as two warriors, identified here as such with their distinctive hairstyle, lie at the base of the hill, bleeding from their necks, wrists, knees, and ankles; their eyes are closed in death. It is the path of footprints that pushes the narrative forward in time.

The endpoint of the journey comes as the Aztecs found what will become their capital city at Tenochtitlan, seen in figure 4.6. Here, seven elders gather

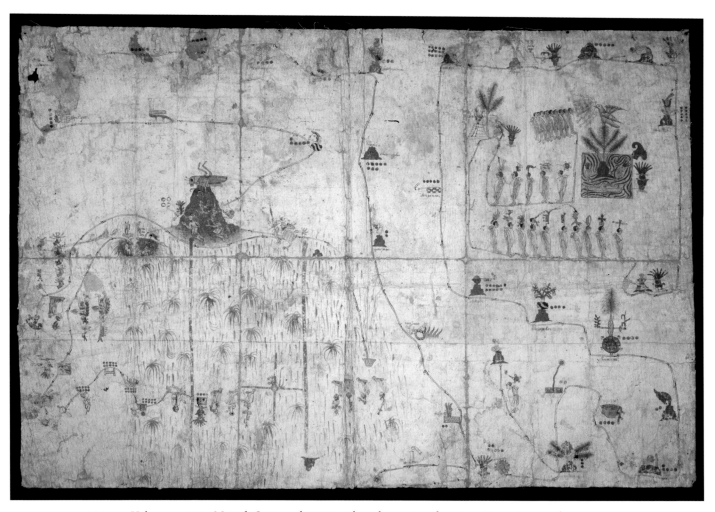

FIG. 4.3 Unknown creator, Mapa de Sigüenza, late sixteenth–early seventeenth century. Pigment on amatl paper, 54.5 x 77.5 cm. Biblioteca Nacional de Antropología, Mexico, 35-14. Artwork in the public domain, reproduction authorized by Instituto Nacional de Antropología, Mexico.

around a blue cross, representing a union of waterways; at the center, a nopal cactus grows. This narrative is a particularly famous one, and finds expression in a number of texts. They recount that the end of the long migration out of Aztlan, which was commanded by their ethnic deity Huizilopochtli, came also at his behest. He told the elders that he would send them a sign that would convey to them the right place for a new settlement, putting an end to their wanderings. In most versions, that sign came in the form of an eagle, a bird with solar associations and thus linked to Huitzilopochtli, a solar deity. The eagle was to alight on a cactus growing in a place near two crossing streams. This would be the site of Tenochtitlan, today's Mexico City. In this version, the eagle does not appear, but the cactus and the crossing streams, as well as the elders as witnesses, do.

While the geographic and temporal span of the journey from Aztlan has never been fully understood, the Aztecs' travels around the Valley in the thirteenth

FIG. 4.4 Detail of figure 4.3.

FIG. 4.5 Detail of figure 4.3.

FIG. 4.6 Detail of figure 4.3, inverted for clarity.

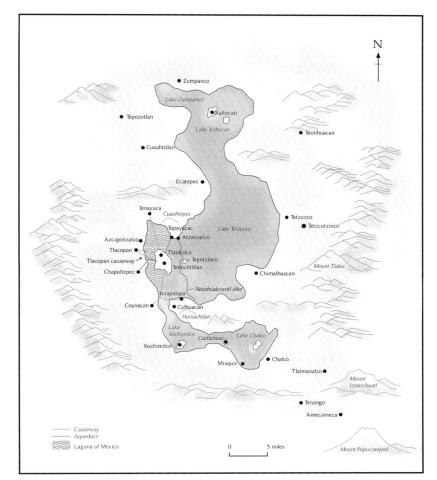

FIG. 4.7 Map of the Valley of Mexico by Olga Vanegas. With permission.

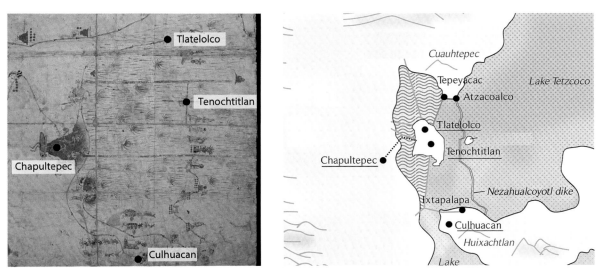

FIG. 4.8 Comparison of locations on Mapa de Sigüenza, figure 4.3, rotated to a northern orientation (left) with detail of figure 4.7 (right). Author diagram.

and fourteenth centuries are better known (fig. 4.7). The lower left of the manuscript, where the Aztecs enter the Valley of Mexico, presents places in roughly cartographic relationships, set along a western orientation; highlighted in the contemporary map in figure 4.8 are Tenochtitlan, Tlatelolco, Chapultepec, and Culhuacan, key places in the narrative. While not meant to show exact scale, the place names are carefully oriented, while the reeds and canals shown in this quadrant of the map offer an approximation of the Valley's distinct ecology.

What was the function of such a map? While it is difficult to reconstruct past viewing or narrative practices, physical and textual evidence is suggestive. The Mapa de Sigüenza is not particularly large, but other examples of these cartographic histories are sizeable. In the Mixtec-speaking region of Oaxaca, they are painted on large cloth sheets, and the wear at the corners of surviving examples shows that they were frequently hung from the corners. Both size and wear suggest a practice of public viewing, with different paths of footprints connecting the distinct narrative threads that are represented, and calendar dates, embedded in the sheet, showing the temporal span and relation of one narrative to another.[20] Ethnographers, working in the field since the nineteenth century, have found knowledge of origins—like those recorded in the sixteenth century in texts like the "Hystoyre"—to be perdurable in Mesoamerica, often being known by indigenous informants.[21] Such imperishability suggests that origin stories were frequently shared and passed down through generations. Shared pasts are a constituting feature of present collectives. So these narrative cartographies—like other history traditions of longue durée across the world—likely functioned to instruct and reaffirm the shared past of a particular people, providing a clear rationale for the occupation of a particular territory while emphasizing the unity of space and time. As public documents, they seem to have been meant to be openly displayed and discussed, updated and copied as needed. They thus offered to the residents of an altepetl the history of that community and the originating rights to a particular place.

World-Making through Conquest

If the migration history offered a rationale for a place to a people, another subset of historical cartography offered the history of a state. Most polities in Mesoamerica were expansionist, in a near-constant state of warfare to dominate other polities and command more resources. As a result, one branch of cartographic histories developed to show the growth of a polity over time, to document successful campaigns of conquest. These images used the same visual language as the community histories.

An example is a page from the *Historia Tolteca-Chichimeca*, a book compiled in the 1560s that lay out the history of a state, beginning with a cave origin.[22] Told via pages of alphabetic texts written in Nahuatl with accompanying images, the book includes several large maps that document the conquests of the Chichimeca Nonohualca, as carried out under two leaders, and the subsequent walking of boundaries to establish a territory. In the map found on folio

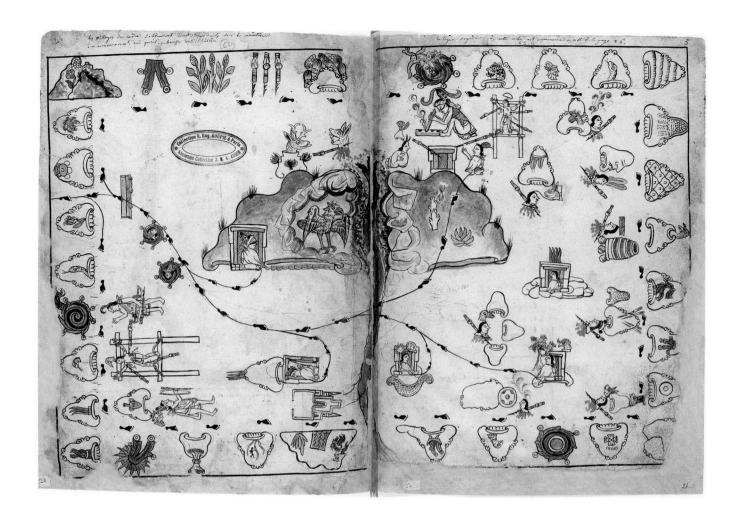

32v–33r (fig. 4.9), the region fills the whole page, ringed by name glyphs of landscape features. Many incorporate the arched hill glyph, which creates the impression of a scalloped border along the edges of the page. Each place, however, is distinguished by a unique pictographic place name. Some of the named places are preexisting settlements; these are usually distinguished by a palace structure, depicted as a white building with a yellow post-and-lintel doorway, seen either frontally or in side view. Seated rulers occupy each palace. A trail of footprints begins at the middle of the left border of the page, to show us two major narrative pathways. One path—a single line of footprints—marches around the edge of the page to show the ritual of walking the boundaries, or surveying the territory. The second is established by footprints that begin as a single set at the left but then separate into five strands, like branches from a tree trunk, to show us the different trajectories of five individual men: founders of the altepetl. Each leader can be followed through the central field to where he eventually takes charge of one of the subsections of the altepetl, an event usually signaled by the assumption of a seat within a palace structure.

The center of the page is dominated by two adjacent hill (*tepetl* in Nahuatl) glyphs, the one at the left containing an eagle (*cuauhtli* in Nahuatl) to create the place name "Cuauhtinchan," which

FIG. 4.9 Unknown creator, Cuauhtinchan foundation scene, Historia Tolteca-Chichimeca, fol. 32v–33r. Pigment on European paper, ca. 27.5 x 40.5 cm. Ms. 46-50, Bibliothèque Nationale de France. Artwork in the public domain, photograph courtesy of the Bibliothèque Nationale de France.

roughly translates to "At the Home of the Eagle." At the bottom left, two Cuauhtinchan leaders, holding bows and arrows, have finished up a sacrifice of a conquered ruler by arrows; another similar sacrifice can be seen at the top right. Other political killings have taken place, as attested by the ten decapitated heads of polity leaders clustered on the right side of the map, each of them linked to a named place. As a result of these conquests of preexisting settlements, which would have involved (and led to the deaths of) many people, the founders could claim that the altepetl of Cuauhtinchan was theirs by right. Notably, no dates are present in this landscape. The lack of calendric notations is explained by the seated figure on the right-hand page, above the right lobe of the central hill glyph. He clasps in his hand a short stick, a fire drill, and is busy drilling it into a wooden board in order to create flames. This action of fire drilling established a new calendar—one of the first actions, after conquest, of foundation. With the successful drilling of fire following the establishment of a determined territory, the altepetl could begin to exist in both space and time.

The vivid display of the executed rulers with blood running from their necks, and, in other cases, shown as being killed by arrow sacrifice, is significant. The depiction of territorial conquest *as* sacrifice served to align the actions of the leaders with the much earlier, and fundamental, sacrifice of the goddess by the creator deities Tezcatlipoca and Ehecatl in order to create the earth. Thus the cartographic history cast human events (the conquest of a particular territory) according to an established template of world creation. In so doing, it also may have served as a guide for future wars and territorial expansion.

Did the cartographic conquest history grow from the migration history? What were the respective origins of these two related genres? These questions are difficult to answer, given the incomplete record of manuscripts that survive. But Mesoamerica also boasts a long history of public monuments, carved stone tablets set in plazas, showing triumphant kings towering over or stepping upon humiliated captives. The presence of similar iconographies, set in a more durable medium than paper or cloth, suggests that the conquest narratives we see in cartographic histories were part of a temporally deep and spatially expansive use of dated conquest imagery deployed by different Mesoamerican states across the centuries. When carved into stone monuments and flanked by calendar dates, they anchored a particular history of conquest to a particular site. When painted on a portable material like cloth or paper, by contrast, they could be used to broadcast the history of territorial conquest widely, perhaps extending their message into the conquered polity or, even more importantly, serving as a template for future action as altepeme jockeyed for preeminence with each other.

The Codex Mendoza as a Hybrid Cartographic History

While the Mapa de Sigüenza and the *Historia Tolteca-Chichimeca* show that the migration history and the conquest history were quite separate variants of the cartographic history, another important manuscript,

FIG. 4.10 Unknown creator, the foundation of Tenochtitlan, Codex Mendoza, fol. 2r, ca. 1542. Pigment on European paper, ca. 21 x 31 cm. Bodleian Libraries, University of Oxford, Ms. Arch. Selden A1, fol. 2r. Artwork in the public domain, photograph courtesy of the Bodleian Libraries.

the Codex Mendoza, reveals a hybrid of the two (fig. 4.10). Through it, we can see how mapmakers and manuscript painters creatively adapted traditional pictorial languages, exhibiting a willingness to experiment with prototypes or genres. The extraordinary Codex was created ca. 1545 by indigenous Aztec artists working in Mexico City after the conquest.[23] The book's given name comes from Antonio de Mendoza, the viceroy of New Spain (1535–1550), and it was almost certainly created for a Spanish patron, or as a gift to a Spanish person, although whether Mendoza or another, we do not know. Its texts are in Spanish, whereas the pictographic language used throughout the manuscript derives directly from the script used by pre-Hispanic Aztec scribes.

Folio 2r functions as a kind of frontispiece to the contents of the manuscript as a whole (fig. 4.10). It also brings together the two cartographic genres (migration and conquest) laid out above, along with a calendar. Painted on European paper by indigenous artists whose names are unknown to us, the standard-sized page is framed by a calendar of fifty-one years, each in a turquoise-colored square. The count begins in the upper left corner and proceeds in a counterclockwise direction, just like the calendar recorded in the "Hystoyre du Mechique." Unlike that count, which begins with 1 Rabbit, or the count in the Tovar Calendar, which begins with 1 Reed, the Mendoza years begin with 2 House (which correlates to 1325 CE), followed by 3 Rabbit, 4 Reed, 5 Flint Knife, and so on, until they reach 13 Reed at the top of the page, a count of fifty-one years. The length of this calendar corresponds to the fifty-one-year reign of Tenoch, one of the elders seen in the center of the page, serving both to synchronize his life span to the movements of the sun, and also to set the history of Tenochtitlan within the count of calendar time.

Within this calendrical frame is set a cartographic space. A blue rectangle traversed by an X offers a schematic view of Tenochtitlan. As seen in figure 4.9, Tenochtitlan was an island city. The blue band of the Codex Mendoza has an undulating edge, a formal characteristic of bodies of water, like the color, allowing us to read the rectangle as a highly abstracted form of the surrounding lake. The X shape within represents crossing streams; we have seen the same crossing streams used to represent Tenochtitlan in the Mapa de Sigüenza (fig. 4.8). Scattered over the field defined by the framing lake are small green plants representing reeds, a typical wetland plant, similar to those in the Mapa de Sigüenza. In the upper quadrant, a small thatched-roof structure is shown in elevation. In the right quadrant, a rack with one pinioned skull represents a *tzompantli*, or skull-rack, a defining feature of ceremonial spaces.

Within the quadrants formed by the arms of the X, ten native leaders appear, each one seated on a small woven-reed seat and each identified with a hieroglyphic name attached to his head. Three of the quadrants contain two leaders; the one at the left contains four. One of them is identified as "Tenuch" (also spelled Tenoch) both by the alphabetic gloss and by his hieroglyphic name. The latter combines a brown-and-ocher oval form, the sign for *tetl*, or stone, with the icon of a nopal cactus, *nochtli* in Nahuatl. Together, these spell out *te-noch*.

While the calendrical frame of fifty-one years expresses an expanse of time, a single historical moment is conveyed within the space of the city, defined by the rectangular lake, where the ten leaders gather. They all face inward, as if to witness the event that transpires at the center of the X. Here is a more elaborate version of the scene that is also shown on the Mapa de Sigüenza. This version includes the eagle, the avatar of Huitzilopochtli, and assigns it the date of 2 House, or 1325. Where the Mapa de Sigüenza adheres closely to the established genre conventions of the cartographic history, relating the whole migration into the Valley of Mexico, the single page of the Codex Mendoza presents only the endpoint of the long journey.

If the narrative of one historical moment—the moment that brought the migration to a close—is captured on the upper part of the page, a more diachronic narrative is found in the page's lower register. Here are displayed two important conquests of other altepeme by the altepetl of Tenochtitlan. They are shown in compressed form: two warriors representing Tenochtitlan each take an enemy captive, the one at the left wielding the *atlatl*, a bladed club, and both wearing the distinctive warrior topknot. Their captives are smaller, with bent knees, as if unable to stand on their own. Each vanquishing warrior faces a burning temple, behind which is the hieroglyphic name of the conquered altepetl. The historical sources do not identify these two altepeme as the first to be conquered by the newly founded Tenochtitlan, although they would be eventually. Instead, the artist's choice to set these places, Culhuacan and Tenayuca, on the page seems to have followed another logic. Both once controlled large tributary empires: Culhuacan, which lay to the south of Tenochtitlan, was an ancient altepetl whose ruling line descended from the Toltecs, an important early empire in the Valley of Mexico; Tenayuca was an earlier capital of an ethnic group known as the Chichimecs, and the seat of their once-expansive empire. Ever conscious of Tenochtitlan's place in history, the artists of the Codex Mendoza seem to have chosen these two conquered places in order to situate their altepetl as the conquerors and successors of states with their own long and illustrious historical trajectory.

During the course of its nearly 200-year history, Tenochtitlan would go on to conquer many other altepeme. The long list of its conquests, all shown as burning temples attached to a hieroglyphic name, fills the pages to folio 16v that follow this one. The two seen on folio 2r thus function as a kind of compressed history, a summary of highlights of what will unfold.

The Way of Telling

It is worth pausing to examine exactly how the narratives are told and the way that space is conveyed in the works presented on the previous pages. The presentation of ideas, be they in the form of images or hieroglyphs with a more restricted semantic meaning, followed highly standardized formats. Bodies are depicted using a frameline to define the body parts, and color is applied evenly, either to fill a space with a uniform color or, at times, as overlapping bands of

different tones of the same color. The head-to-body proportion is variable, but usually consistent within a single manuscript. Human figures are shown in profile and exhibit a very limited range of gestures and positions. As many scholars have noted, the principal goal of this pictorial script is for a clear presentation of events (Elizabeth Boone likens it to *res gestae*), with little attention paid to conveying the emotional register of actors, or even their motivations.[24]

In many respects, this pictorial script was a perfect vehicle for cartography. While place names were rendered hieroglyphically, some parts of the sign were rendered with natural referents in mind. The *tepetl* ("hill") glyph takes the form of an inverted "u," flaring at the base, and it is frequently painted green. Many place names include the "*tepe*" root and thus deploy the *tepetl* glyph; at other times, a *tepetl* is included in a place name to convey the idea of "altepetl" rather than any semantic content. Since the surfaces of manuscripts did not need to present alphabetic texts and order them in accord with a reading logic, artists were free to arrange glyphs across the surface of the page to mirror the arrangement of places in space. The surface of the support, be it cloth, paper, or sized deerhide, could thus function as an analogue of the surface of the earth.

Inasmuch as the Mapa de Sigüenza represents the marshy landscape surrounding Tenochtitlan (fig. 4.6), its creator almost certainly had seen European landscapes and was adapting from them, for it is rare to find landscape elements that lack semantic or symbolic content represented in pre-Hispanic manuscripts.[25] Aztec maps show a marked tendency to create a landscape out of names, rather than through naturalistic features lacking symbolic import, unlike the caves and hills that were features of the deity Tlaltecuhtli's body. In order to represent a place on a page, that place needed to be created, or elevated to the level of representation, by being given a name. But where did names come from? While Mesoamerican accounts of town foundation specify the walking of boundaries, and the shooting of arrows to four directions to define a communal space, they do not specify the invention of a name by town founders. Names seem to preexist foundation, or to have been generated through foundation. Consider the scenario that we see taking place on the Codex Mendoza folio 2r (fig. 4.10). Here, the name "Tenochtitlan" derives from *tetl* (rock), *nochtli* (nopal cactus), the ligature *ti*, and *tlan*, "next to." The rock and nopal cactus may have been landscape features, but they were more importantly part of Huitzilopochtli's sign to the wandering Aztecs that this was the site for settlement. Thus, names were not a result of human choice, but often the commemoration of divine actions that shaped the world.

Although the tepetl glyph often serves as part of a place name, a close examination of its iconography points to the complicated fashion in which hieroglyphs could both have a linguistic function as a sign (standing for the morpheme "tepe") as well as carry a more ample symbolic meaning. In some cases, they connect the representation back to the fundamental actions of world creation that we know from manuscripts like the "Hystoyre." On a map from the Misantla region of Veracruz, painted by indigenous

FIG. 4.11 Unknown creator, land map from Zolipa, Misantla, Veracruz, 1573. Tierras 2672, 2nd part, exp. 18, fol. 13. Archivo General de la Nación, Mexico, Fondo MAPILU: Mapas, Planos e Ilustraciones, Núm. 1535.

FIG. 4.12 Unknown creator, place name of "Xico, pu[ebl]o," after Codex Mendoza, fol. 20r, detail, ca. 1542. Author drawing.

artists in 1573 (fig. 4.11), for instance, the two mountains at the bottom left and upper right of the town are shown with the familiar tepetl glyph, seen close-up and slightly cropped by the edge of the page. The mapmaker has chosen, however, to mark the surface with a particular diamond-and-dot pattern, a pattern seen on other maps as well. In the corpus of Aztec stone sculptures, this pattern is used to depict the skin of Tlaltecuhtli, the same creature whose ripped-apart body gave rise to the earth in the "Hystoyre." Mapmakers were thus recalling the divine origin of the earth's features. Two other hieroglyphs reveal more about the meaning of the *tepetl* glyph, and the hills that they stood for. The Codex Mendoza, whose map of Tenochtitlan was discussed above, also contains the names of the polities conquered by the Aztec, expressed hieroglyphically. The alphabetic annotations, added to the manuscript almost immediately after it was painted, have aided immensely in the understanding of how the hieroglyphs were composed. The place name "Xico" derives from the Nahuatl *xictli* ("navel") and the locative *–co* ("place of"). The hieroglyph on Codex Mendoza folio 20v shows us a navel with a ropelike umbilical cord extending from it (fig. 4.12). Significantly, the yellow circle containing the navel is surrounded by a red ring and then a green ring marked with foliate "stony" symbols. Through color and form, the artist thus conveys that we are looking at the underside of the *tepetl* glyph; since only mammals have umbilical cords, the iconography suggests that Tlaltecuhtli was a kind of mammal, and that hills, and the *tepetl* glyphs that represented them, were once part of that mammalian body.

So even a hieroglyph made reference to narratives of world creation, allowing us to see that Aztec maps brought together in different ways historical narratives of different scale, be they a cosmography from deep time or the marking of community boundaries and the establishment of calendars by known human actors. Alfredo López Austin's theory that Mesoamericans shared a holistic and integrative worldview, which he calls *cosmovisión*, helps us understand that these scales—of the divine, of the human—were mutually constitutive and reinforcing. In particular, the bloodshed and violence required in the *Historia Tolteca-Chichimeca* to create the territory of Cuauhtinchan was the echo of equally productive acts of cosmic violence. "The basis of cosmovision," López Austin writes, "was not just the product of speculation, but derived from practical and daily relationships; it grew out of a determinate perception of the world, was conditioned by traditions that guided human actions in relation to society and in relation to nature."[26] Thus, narratives on maps in an Aztec or Mesoamerican context are best understood in relation to an integrating vision of human beings within the larger cosmos, itself conceived as movement, the defining feature that united space and time into a single ontological cluster.

Most of the examples we know of this tradition come from the sixteenth century, many of them painted after the Spanish conquest. In the map of Zolipa, for instance, the town at the bottom (within the darker water stain) is shown with a chapel or small church, identifiable as such by the small cross appearing on the thatched roof (fig. 4.11). This is a land-

grant map, commissioned by a Spanish petitioner seeking to be given land by the crown—one sign of the vast transfer of lands out of indigenous control that would begin in the sixteenth century and that continues today. Likely a seventeenth-century copy of an earlier map, the Mapa de Sigüenza was made at a time when the Spanish were firmly in control of the lands depicted (fig. 4.3). Despite this, indigenous mapmakers, and often their patrons, still saw the need to create and copy maps that offered key historical narratives justifying possession. In so doing, they reaffirmed distinct worldviews about the nature of space and time during a period of dramatic change brought by the colonial regime.

Afterlife

The Mapa de Sigüenza was one of the most frequently reproduced Aztec maps of the eighteenth and nineteenth centuries. And, like many monuments surviving over time, its interpretation served the needs of the present. Looking at its more recent history shows us not only how the Aztecs were slotted into different historiographic frames, but also, more importantly, how the emergence of new spaces, like the nation, called for new reckonings of time. In 1780, picking up on a historical trope established in the sixteenth century, the Jesuit Francisco Javier Clavigero claimed that the Mapa de Sigüenza showed a creation myth of the world after the flood, thereby demonstrating the parallel between the history of ancient Mexicans and that told in the book of Genesis.[27]

But when a lithographic version of the map appeared in the mid-nineteenth century in the first national atlas of Mexico in 1858, it was used to establish a different chronotope, one concerned with the beginnings of the nation. Antonio García Cubas's *Atlas geográfico, estadistico é historico de la República Mexicana* is notable for appearing at a charged moment in Mexico's history.[28] After its successful war of independence from Spain, ending in 1821, Mexico had been battered by setbacks: the Mexican-American War of 1846–1848 and the Gadsden Purchase of 1854 resulted in the loss or sale of over half of Mexico's lands to the United States. Published a mere four years after the dramatic truncation of the national territory, the *Atlas geográfico* opened with a national map showing the reconfigured territory; around it, the nation's vital statistics were arrayed through texts and graphs—the number of inhabitants, their racial makeup. The new northern frontier could be seen as a bright red line across the top of the map, identified as "Limit between Mexico and the United States, following the Treaty of La Mesilla [and] Guadalupe, signed on the 2nd of February 1848." The *Atlas* was certainly a work of national propaganda for viewers inside and outside Mexico. As such, it placed particular focus on Mexico's healthy economic production, with the first folio featuring tables that registered the quantities of the products that fed the export market, such as precious metals, sugar, and cochineal dye. García Cubas's aim to address the lack of statistical and geographical information about Mexico was a leitmotiv, and the 1858 publication was the first of five important atlases he made during his long career, which included the publication of geographic

FIG. 4.13 Mapa de Sigüenza, lithographic copy. From Antonio García Cubas, *Atlas geográfico, estadistico é historico de la República Mexicana* (Mexico: José Mariano Fernandez de Lara, 1858), sheet 28, detail. Artwork in the public domain, photograph courtesy of the David Rumsey Map Collection, https://www.davidrumsey.com/.

and statistical information in Spanish, French, and English, as well as the first geographic treatise for primary school teachers.[29]

Set at the end of the national atlas, the bifolio that contained the color lithographic version of the Mapa de Sigüenza was formatted similarly to the ones featuring modern maps that preceded it (fig. 4.13). The map image dominates the center of the page and is framed by typeset texts. The visual harmony smooths over the oddness of this map—a map that bore witness to the country's Aztec (and pagan) past, within the context of an *Atlas* that was, as a whole, designed to highlight the nation's participation in progressive modernity. Yet including the ancient Aztec map was very much in keeping with an emergent narrative about the nation's historical origins. Mexico's leaders had chosen the Aztec past above other possible points of national origin, preferring it to the Spanish

History in Maps from the Aztec Empire

conquest, which had been the conventional point of origin during the colonial (or Viceregal) period. Subsequent to Mexico's independence movements of 1810–1820, the idea of locating the nation's origin with the Conquest of 1519–1521, the moment when Mexico came under the control of the Spanish crown, was less than desirable. By contrast, the military prowess of the Aztecs and their expansionist policies were attractive qualities to highlight in a national history when leaders were charged with rebuilding the country following the disastrous war with the United States. Moreover, the Aztec map showed the ancient peregrination leading to Tenochtitlan, which lay under Mexico City, the nation's indisputable political and cultural capital. García Cubas's lithographic rendering, colored in yellow, green, red, light brown, gray, and blue, made the narrative easier to see and easier to read. Numbers accompanied each of the narrative episodes, which were keyed to explanatory texts at the side.

At the same time that the presence of the Mapa de Sigüenza in the national atlas anchored the nation of Mexico and its capital city to an Aztec point of origin, the texts that framed this "pagan" map also reasserted a fundamentally Christian cosmology: Mexico was in the nineteenth century, and still is today, a predominantly Roman Catholic nation. To write the texts, García Cubas asked José Fernando Ramirez, who was the curator of the National Museum (founded in 1825, it was another national project), which housed, and still houses, the largest corpus of Aztec sculpture in the world. Drawing the reader's attention to the visible trails of footprints on the surface of the image, Ramirez emphasized that the map offered "a trail left by the primitive traditions of the human race, which linked the Americas with Asia, and tied together all the races disseminated over the globe, connecting them back to the first family of creation."[30] Designed to show the interrelation of space and time in the Aztec world, the Mapa de Sigüenza served, by the nineteenth century, to synchronize the origins of the Mexican nation with both the origins of the Aztecs and the deeper time of the Bible, revealing the ongoing interrelation between the map and the calendar, and between space and time.

Notes

1. Norbert Elias, *An Essay on Time*, ed. Steven Loyal and Stephen Mennell (Dublin: University College Dublin Press, 2007), 18.

2. Elias, *An Essay on Time*, 60.

3. Edward Casey, *Getting Back into Place: Toward a Renewed Understanding of the Place-World*, 2nd ed. (Bloomington: Indiana University Press, 2010).

4. See also Henri Lefebvre, *The Production of Space*, trans. Donald Nicholson-Smith (Malden, MA; Oxford: Blackwell, 2009). Elias, *An Essay on Time*, 80–83, also discusses the relation of time and space.

5. The concept comes from François Hartog, *Regimes of Historicity: Presentism and Experiences of Time*, trans. Saskia Brown (New York: Columbia University Press, 2015). It is discussed in the Aztec context by Federico Navarrete Linares, "Writing, Images and Time-Space in Aztec Monuments and Books," in *Their Way of Writing: Scripts, Signs, and Pictographies in Pre-Columbian America*, ed. Elizabeth Hill Boone and Gary Urton (Washington, DC: Dumbarton Oaks Research Library and Collection, 2011), 176.

6. Catalogued as Français 19031, "Fragments d'André Thevet

sur les Indes occidentales et sur le Mexique." A transcription of the manuscript was published with an introduction in Edouard de Jonghe, "Histoyre du Mechique, manuscrit français inédit du XVIe siècle," *Journal de la Société des Américanistes* 2, no. 1 (1905): 1–41. Jonghe believed the BNF manuscript to have been transcribed and translated by the French Franciscan explorer André Thevet (1516–1590). He posits that the original Spanish text was written by the Franciscan Andrés de Olmos (ca. 1485–1571) and entitled *Antiguedades mexicanos*. The manuscript can be consulted online at https://gallica.bnf.fr/ark:/12148/btv1b9062312t/.

7. Jonghe, "Histoyre du Mechique," 29.

8. The source spells it "Tlalteuhtli," but I have used the more conventional spelling here.

9. Jonghe, "Histoyre du Mechique," 29.

10. Jonghe, "Histoyre du Mechique," 30.

11. This temporal principle has been expressed by the philosopher James Maffie as "*olin* motion change" (with *olin* being the word in Nahuatl, the Aztec language, for movement). Maffie, *Aztec Philosophy: Understanding a World in Motion* (Boulder: University of Colorado Press, 2014), 185–260.

12. The thesis that human sacrifice "feeds" the sun was articulated by Alfonso Caso in 1927 and remains the generally accepted rationale for the practice. Caso, *El teocalli de la guerra sagrada: (descripción y estudio del monolito encontrado en los cimientos del Palacio Nacional)* (Mexico City: Departamento de Educación Pública, Talleres Gráficos de la Nación, 1927).

13. Anthony Aveni surveys known calendars and argues that pre-Hispanic calendars were conceived of as square, and the round form may have been influenced by European calendars. See *Circling the Square: How the Conquest Altered the Shape of Time in Mesoamerica* (Philadelphia: American Philosophical Society, 2012).

14. A lucid introduction to the Mesoamerican calendar is to be found in Anthony F. Aveni, "Timely Themes: An Introduction to the Measure and Meaning of Time in Mesoamerica and the Andes," in Aveni, ed., *The Measure and Meaning of Time in Mesoamerica and the Andes* (Washington, DC: Dumbarton Oaks Research Library and Collection, 2015), 1–8.

15. The text reads: Ce Tochtli / Midi / Vitztla; Ome Acatl / Orient / Tlapco; Yey Tecpatl / Septentrio / Teotlalpa; Naui Calli / Occide̅s / Ciuatla. I have normalized the spelling of Nahuatl words in the main text.

16. The earliest historians of indigenous Mexican cartography were quick to note the connection, starting with C. A. Burland, "The Map as a Vehicle for Mexican History," *Imago Mundi* 15 (1960): 11–18. See also Elizabeth Hill Boone, *Stories in Red and Black: Pictorial Histories of the Aztecs and Mixtecs* (Austin: University of Texas Press, 2000); and Barbara E. Mundy, "Mesoamerican Cartography," in *The History of Cartography*, vol. 2, bk. 3, *Cartography in the Traditional African, American, Arctic, Australian, and Pacific Societies*, ed. G. Malcolm Lewis and David Woodward (Chicago: University of Chicago Press, 1998), 183–256, for more recent discussions of the cartographic history.

17. María Casteñada de la Paz, *El mapa de Sigüenza:* una nueva interpretación de la Pintura de la peregrinación de los Culhua-Mexitin (México: CONACULTA-Instituto Nacional de Antropología e Historia, 2007).

18. Different sources give different time spans for the migration. See Federico Navarrete Linares, *Los orígenes de los pueblos indígenas del Valle de México: los altépetl y sus historias* (México: Universidad Nacional Autónoma de México, 2011).

19. Michael E. Smith, "The Aztlan Migrations of Nahuatl Chronicles: Myth or History?," *Ethnohistory* 31, no. 3 (1984): 153–86.

20. Barbara E. Mundy, "At Home in the World: Mixtec Elites and the Teozacoalco Map-Genealogy," in *Painted Books and Indigenous Knowledge in Mesoamerica: Manuscript Studies in Honor of Mary Elizabeth Smith*, ed. Elizabeth Hill Boone (New Orleans: Middle American Research Institute, 2005), 363–82.

21. For an example of a study that connects contemporary knowledge, gained through ethnography, to the content of fifteenth-century manuscripts, see Maarten E. R. G. N. Jansen and Gabina Aurora Pérez Jiménez, *Encounter with the Plumed Serpent: Drama and Power in the Heart of Mesoamerica* (Boulder: University Press of Colorado, 2007).

22. Dana Leibsohn, *Script and Glyph: Pre-Hispanic History,*

Colonial Bookmaking and the Historia Tolteca-Chichimeca (Washington, DC; [Cambridge, MA]: Dumbarton Oaks Research Library and Collection; distributed by Harvard University Press, 2009); and Paul Kirchhoff, Lina Odena Gűemes, and Luis Reyes García, eds., *Historia Tolteca-Chichimeca* (México: Instituto Nacional de Antropología e Historia, 1976).

23. Frances Berdan and Patricia Rieff Anawalt, eds., *The Codex Mendoza*, 4 vols. (Berkeley: University of California Press, 1992).

24. Boone, *Stories in Red and Black*, 10.

25. See the discussion of Codex Nuttall in Mundy, "Mesoamerican Cartography," and John Pohl and Bruce Byland, "Mixtec Landscape Perception and Archaeological Settlement Patterns," *Ancient Mesoamerica* 1, no. 1 (1990): 113–31; and Manuel A. Hermann Lejarazu, "El sitio de Monte Negro como lugar de origen y la fundación prehispánica de Tilantongo en los códices Mixtecos," *Estudios Mesoamericanos,* Nueva época 10, Jan.–Feb. (2011): 39–61.

26. Alfredo López Austin, *Tamoanchan y Tlalocan* (México: Fondo de Cultura Económica, 1994), 15: "Aún más: la base de la cosmovisión no es producto de la especulación, sino de las relaciones prácticas y cotidianas; se va construyendo a partir de determinada percepción del mundo, condicionada por una tradición que guía el actuar humano en la sociedad y en la naturaleza."

27. Quoted in José Fernando Ramirez, "Cuadro historico-geroglifico, las tribus Aztecas I," in Antonio García Cubas, *Atlas geográfico, estadistico é historico de la República Mexicana* (Mexico: José Mariano Fernandez de Lara, 1858), sheet 28.

28. García Cubas, *Atlas geográfico*.

29. Antonio García Cubas, *Compendio de geografía universal para uso de los establecimientos de instrucción primaria* (México: Murguía, 1909). On García Cubas's role in nineteenth-century cartography, see Raymond B. Craib, *Cartographic Mexico: A History of State Fixations and Fugitive Landscapes* (Durham, NC: Duke University Press, 2004). His atlases are also discussed in Magali Marie Carrera, *Traveling from New Spain to Mexico: Mapping Practices of Nineteenth-Century Mexico* (Durham, NC: Duke University Press, 2011).

30. The Spanish original reads: "un rastro de las tradiciones primitivas del género humano, que enlazan la América con el Asia, y eslabonan todas las razas diseminadas por el globo, reduciéndolas á la primera familia de la creacion." García Cubas, *Atlas geográfico*, sheet 28.

5

Veronica Della Dora

Lifting the Veil of Time

Maps, Metaphor, and Antiquarianism in the Seventeenth and Eighteenth Centuries

And when time's veil is rent away,
Whereby eternity is hid,
When though shalt all things open lay,
Which ere we thought, or said, or did,
Among time's ruins bury so
Our failings through our tract of time
That from these dungeons here below
We to celestial thrones may clime,
And there to our eternal king,
For ever Hallelujah sing.
—GEORGE WITHERS, For the Day Present, or the Last Day (1622)

Maps outline futures and make pasts visible. As projects on territory, they give visual expression to what has yet to come; as historical records, they open windows onto bygone worlds. Urban plans and maps of territorial partitions have the power to shape the land, or simply to conjure up new spatial orders in our mind. Historical maps, for their part, help translate time into space. As Ortelius wrote, *historiae oculus geographia*: "geography allows history to be visualized on terra firma."[1] In pinning down locations and events of the past, historical maps bring the past before our eyes. But what about the passing of time? And what about *time* itself? Can time be represented at all?

In a way, time seems to be antithetical to cartographic representation. The map, Christian Jacob argues, is "an objective image that a society uses to set in place the figure of the world and its totality, that is, of what preexists, of what outlives the individual who is its ephemeral passenger. The synoptic gaze requires an absolute synchronism; the history comes to its end on the frozen image of the world."[2] Mapping is first of all a cognitive process through which we make the world legible to ourselves. It is a process through

which we order, tame, and crystallize reality in order to make it comprehensible, if only to our mind. Time, by contrast, is by definition fluid and elusive. While it can be measured, it can never be pinned down; it seems to resist representation. It escapes our grasp. As Basil of Caesarea wrote in the fourth century, "Is not this the nature of time, where the past is no more, the future does not exist, and the present escapes before being recognized?"[3]

Time flows, time flies, time runs, time unfolds. Time steals and heals. Time reveals and conceals. As with all ungraspable things—and perhaps more than any other—we can speak of time only through metaphors. Metaphors, cognitive linguistics teaches us, are not just poetic flourishes; they are the principal way through which we make sense of abstract concepts; they are a part of human thought. Metaphors project an image, or a pattern, from a familiar (and usually more concrete) domain of experience onto another unfamiliar (and usually more abstract) domain, in order to illuminate meaning on something included in the latter domain.[4] The more abstract and elusive something is perceived to be, the larger the number of metaphors clustered around it. Time falls in this category. Because we cannot directly sense time, we can only explain our relationship to it by way of metaphorical language. Augustine, for example, spoke of the progression of the centuries as a great song, the work of an ineffable musician; he thus set the flow of time in relation to musical rhythm.[5] Charles Dickens called time "a spinner." Thoreau likened it to a stream, whose "thin current slides away, but eternity remains."[6]

This chapter addresses this book's theme of space-time interactions on traditional cartographic media by way of one such metaphor: the veil. In particular, it explores visualizations of "time's veil" in western Baroque cartography. During the seventeenth and eighteenth centuries, globes unveiled by personifications of Father Time became common features on the frontispieces of atlases and other works. Serving also as cartouches, page titles, insets, or simple decorations, flying veils, airborne banners, opening curtains, and heavy drapes lifted by winged putti were likewise ubiquitous presences within maps and bird's-eye views. In some cases, maps themselves were represented as drapes or stage curtains. The purpose of these veils was as manifold as their shapes. They could be used to build anticipation, to convey the excitement of unfolding discovery, to trumpet the progress of a military campaign, to celebrate the advancement of knowledge of the world and of the past. In all cases, their function was primarily rhetorical: like the map, the veil conceals and reveals; it is a metaphor for what can be seen and what is hidden from us. The veil is a thin layer that separates the present from the past and the future; it is a membrane that divides but also connects our world and other worlds.

While the veil metaphor has a transcultural appeal, in early modern Europe it nonetheless held a special cultural significance.[7] It was a metaphor not only for the unfolding of time, but also for a specific approach to knowledge at a specific moment in time. Its peak of popularity during the seventeenth and eighteenth centuries can be linked to a taste for

theatricality typical of the Baroque. More significantly, however, it should be connected to the rise of experimental science and the importance placed on discovery, the unveiling of the mysteries of nature. The verb "to unveil" entered the English language as late as the 1590s (with the meaning of "to make clear"), and it acquired the sense of "to display" or "to reveal" in the 1650s.[8] The English philosopher and statesman Francis Bacon talked about the ability of experimental history to "take off the mask and veil from natural objects, which are commonly concealed and obscured under the variety of shapes and external appearance."[9]

More specifically, the veil also gave expression to the dialectics between light and darkness underpinning Baroque cosmographic discourse. Some of the most original uses of "cartographic veils" were pioneered by the main proponents of cosmography of the time, including (and especially) the Venetian Vincenzo Coronelli (1650–1718). Set in this context, the veil is but an expression of an emblematic culture of "metaphysical enlightenment" in which knowledge was at once "hidden and arcane and illuminating."[10] The discovery and interpretation of the *arcana* of the cosmos went hand in hand with the development of antiquarianism from mere collecting practice to the study of the past, as well as with a longer Renaissance tradition of creative appropriation of classical mythology to forge emblems.[11] Unsurprisingly, at this time the ancient motto *veritas filia temporis*, whereby Truth is unveiled by Father Time, became one of the most popular subjects in painting across Europe, which in turn informed the iconography of frontispieces to scientific works, including atlases.

The first part of the chapter explores the genealogy of the "veil of time" metaphor by contextualizing it within this antiquarian tradition. The following section focuses on the representation of Time's veil on the frontispieces of early eighteenth-century Dutch atlases and other geographical works. Home to the greatest merchant fleet in the world and to prestigious universities, and a safe haven for religious refugees from neighboring countries, the Netherlands was the main printing industry and book trade center in Europe from the seventeenth through the mid-eighteenth centuries.[12] The extraordinary output and diversity of Dutch book production is clearly reflected in the quantity and quality of geographical works, whose frontispieces acted as showcases for the intellectual wonders of the Dutch Golden Age.[13] This section of the chapter shows how Time's veils on such frontispieces helped conjure a sense of expectation and a narrative of progress that responded to the contemporary fascination with discovery.

The third section moves from the Netherlands to Venice (another major European printing center), and from the frontispiece to the body of the book. It considers representations of veils within maps and of maps as veils, with a specific focus on Coronelli's work. It shows how cartographic veils—operating together with the dynamic media of travel accounts, atlases, and military propaganda books—were used to reinforce the sense of unfolding of events and their spatialization, to the point that maps themselves became time's veils.

The Veil of Time

Textile metaphors have long been used to express the fluidity of time and its multiple facets. This is because fabric is so malleable. Draped, folded, and inflated, fabric, like time, constantly morphs. It is always the same and yet never the same. As it is spun and woven, folded and unfolded, stored away and unrolled, fabric seems appropriate for representing memory, the recollection of past events, but also the unfolding of the future. Second-century Roman statuary thus portrayed Mnemosyne, the daughter of Chronos (Time) and personification of memory, veiled to the tip of her fingers, whereas in the classical world Hores and Moires were believed to spin human destiny.[14] In Homer the word *kairos* referred to the measuring out at birth of the portion of wool or flax to each person with which their thread of fate was to be made.[15] In Virgil's *Aeneid*, to repeat the story of a fate was to "wind or roll it again" as one winds and unwinds a ball of string.[16] The passing of life thus took the form of a weaving, whose completion marked death, with the fabric being cut off from the loom.[17]

Twelfth-century philosopher and poet Bernardus Silvestris talked about "the fabric of time," whereas Dante likened the world to a mantle cut by the "scissors of time."[18] Identified with Saturn since antiquity, time also took a human shape. In classical art he featured as a dignified but grim figure, his head veiled as he grasped a sickle, an iconography that was occasionally resurrected in the Middle Ages.[19] By the end of the sixteenth century, allegories of Father Time had assumed new shapes and attributes. In his influential *Iconologia* (1593), Cesare Ripa offered different descriptions: Time was always an old man, now winged (from which the saying *volat irreparabile tempus*), now dressed in a mantle of stars, now with his white-haired head covered in a green veil, as to signify the rebirth of spring out of the snows of winter.[20] In Geoffrey Whitney's *Choice of Emblems* (1586), Time was represented as a muscular old man with wings on his back (referring to Time's swiftness) and a scythe in his hand.[21] At other times, his mighty countenance was adorned with the attribute of the clock or the hourglass.[22]

If in antiquity and in the Middle Ages Chronos, like his daughter Mnemosyne, was veiled, in the Renaissance it was his ability to *unveil* that captured the imagination of scholars, artists, and antiquarians. In the adage *tempus omnia revelat* (time reveals everything), Erasmus cites many metaphorical instances of time "revealing truth" or "bringing truth to light" going as far back as Pindar and Sophocles—in other words, what time veils is also what it unveils. Erasmus nonetheless ascribes the motto *veritas filia temporis* to the Roman author Aulus Gellius, who tells us that "some ancient poet called Truth 'the daughter of Time,' because though for a space she may lie hidden, yet with the progress of time she comes forth into the light." Erasmus also explains that "Plutarch in his *Problems* poses the question why it is that the ancients were accustomed to sacrifice to Saturn with their heads covered; and he thinks the point was that truth as a rule is covered and unknown, but is nonetheless revealed by time. For Saturn is fabled to be the creator and god of time." In the midst of these classical citations, Erasmus quotes Scripture and ear-

FIG. 5.1 Agnolo Bronzino, *An Allegory with Venus and Cupid* (1545). © The National Gallery, London.

ly Christian authors such as Tertullian, using classical pagan texts and the Bible together to weave a seamless tapestry of wisdom and moral teaching.[23]

This moral emblematic tradition of time as a revealer finds one of its earliest and most dramatic visual expressions in the silky blue drape brutally lifted by Father Time in the Italian painter Agnolo Bronzino's *An Allegory with Venus and Cupid* (1545) (fig. 5.1). In a subtle play of allegories, winged Time, aided by Truth, unveils the vices and passions hidden behind the allures of love. The veil pulls back to show that lurking behind the soft, sensual allegories of Pleasure and Jest (foreground) are Jealousy (the hair-pulling figure at left) and Deceit (at right, represented by a

FIG. 5.2 Theodoor van Thulden, *Time Revealing Truth* (ca. 1650). Lawrence Steigrad Fine Arts.

girl's head on a sphinxlike body). As the art historian Erwin Panofsky explained, this image shows "the pleasure of love on the one hand and its tortures and dangers on the other, in such a way, however, that the pleasures are revealed as futile and fallacious advantages, whereas the dangers and tortures are shown to be great and real evils."[24]

The exposure of vice through the unveiling of the whole group can be read as a variant of the motto *veritas filia temporis* (truth is the daughter of time), which Bronzino explicitly represented as part of the *Innocentia* tapestry embroidered three years later. As Fritz Saxl and others have shown, throughout the sixteenth century the motto was appropriated and exploited for political and religious goals. Intriguingly,

it was used by English monarchs from Henry VIII to Elizabeth I to serve the causes of Protestantism and Catholicism in turn. At the time of the Reformation, "in less than twenty-five years the emblem had reversed the meaning twice, and had been made the vehicle three times of strong emotion."[25]

Tempus omnia revelat and *veritas filia temporis* nonetheless always retained a broader and more abstract moral connotation and a proverbial significance that transcended politics.[26] For the next two centuries, Father Time rescuing and unveiling naked Truth endured as one of the most popular themes of western European art, from Peter Paul Rubens and Gian Lorenzo Bernini to Pompeo Batoni.[27] Time's veil of fabric—sometimes echoed in dramatic fabrics of clouds—revealed the Baroque fascination with the play between light and shadow, concealment and revelation.

In a painting made around 1650 by Dutch artist Theodoor van Thulden, for example, Time tears away a red and golden cloth to reveal Truth's naked body (fig. 5.2). Truth holds a sun, her traditional attribute, which illuminates her entire body, showing that she has nothing to hide. The title page of a half-opened book tucked under her left arm reads "Sol et Tempus Veritatem Detegunt" (Sun and time reveal truth). On the barren ground beneath her, a mask is transfixed by Time's scythe, symbolizing the lies that are overcome by Time and Truth. A ruined pillar at left in the background reminds the viewer of the passing of time and the transience of human affairs.[28]

Transiency and the victory of time and truth over terrestrial things is a theme expressed with special

FIG. 5.3 Detail from Alexander VII's funerary monument in the Basilica of Saint Peter, Vatican. Photo: Jean-Pol Grandmont, public domain.

intensity by Bernini. In his *Truth Being Unveiled by Time* (1652), Veritas sets her foot on a small globe, signifying her supremacy over mundane concerns (the iconography is also linked to the motto "veritas de terra orta est" [Ps. 85:12]). The action of uncovering is made the main theme of the composition. Veritas's eyes are turned upward, away from the globe and toward the approaching figure of Time (though the figure of Time was never executed). Her stretching arm displays the sun to the world. Unveiling and active detachment from the earth are made to coincide: Veritas, according to Saxl, "longs for the moment of unveiling like a bride."[29] By contrast, in the funerary monument of Alexander VII Chigi (1678), Truth appears alongside the allegories of three other virtues (Charity, Justice, and Prudence), holding the sun tight in her bosom (fig. 5.3).[30] Being eternally manifest, Truth is now by no means revealed by Time. On the contrary, the Four Virtues' gigantic marble veil covers Death and Time for eternity.

Lifting the Veil

This brief excursion into the genealogy of representations of Time and more specifically the motto *veritas filia temporis* helps us understand the popularity and complex iconography of the unveiling motif on Baroque frontispieces. According to Panofsky, "no period has been so obsessed with the depth and width, the horror and the sublimity of the concept of time as the Baroque, the period in which man found himself confronted with the infinite as a quality of the universe instead of as a prerogative of God."[31] The seventeenth-century Scientific Revolution saw the disintegration of the self-enclosed and divinely ordained Aristotelian cosmos that had previously dominated Western thought. Its reassuring finitude was replaced by "an indefinite and even infinite universe bound together by the identity of its fundamental components and laws."[32] At the opposite end of the scale, the invention of the microscope and the pio-

neering of microbiology opened up a totally new and equally unexplored universe.

On seventeenth- and eighteenth-century frontispieces of anatomy and natural philosophy treatises, Truth was often conflated with a self-unveiling Nature, sometimes portrayed as Isis.[33] As the secrets of nature were unveiled and made visible thanks to the microscope and the telescope, truth was brought to light and, in the words of the German astronomer Johannes Kepler, "man became the master of God's works."[34] Time was of course central to this process of unveiling, and its personification made its sublime, ineffable quality perhaps somewhat easier to master. At the same time, however, the dynamic presence of the unfolding veil conveyed the tension between known and unknown embedded in the process of discovery and the infinite, continuously expanding horizons of human knowledge. As the English philosopher Francis Bacon famously claimed in his *Novum Organum* (1620),

> We must also take into our consideration that many objects in nature fit to throw light upon philosophy have been exposed to our view and discovered by means of long voyages and travels, in which our times have abounded. It would indeed be dishonorable to mankind, if the regions of the material globe, the earth, the sea, and stars should be so prodigiously developed and illustrated in our age, and yet the boundaries of the intellectual globe should be confined to the narrow discoveries of the ancients.[35]

Thus understood, knowledge was a process both resting on and challenging previous discoveries. As Bacon indicated, geographical knowledge was no exception; it rather served as a metaphor for the progress of human knowledge in general. Bacon appropriated the ancient motto *veritas filia temporis* to describe the progressive unveiling of Truth by Time and the slow discovery of the secrets of Nature and of the planet, thanks to the collective efforts of mankind.[36] The French mathematician Blaise Pascal later remarked that "the mysteries of nature are hidden, [...but] Time reveals them from era to era."[37]

Progressive revelation informs the iconography of a number of late seventeenth- and eighteenth-century frontispieces and title pages of atlases and other geographical works. Here a gradually unveiled globe takes the place of Nature and Truth. The globe moves from attribute to prominent presence and protagonist. On Joannes Wolters's edition of *Strabonis Geographia cum notis Casauboni et aliorum* (1707) (fig. 5.4), for example, Time's veil acts as a cosmographic stage curtain gradually opening on the theater of the world from the Greek world of Strabo in the first century BCE to the present. Here a winged, scythe-bearing Father Time unveils Europe, Asia, and Africa as young female allegories point to their respective continents. The western hemisphere remains shrouded, a reminder that this part of the world was unknown to Strabo. History patiently records these unfolding events in her book. As with Isis's opening veils on the frontispieces of scientific books, here the veil of Time ultimately unveils truth, reminding the reader of human progress since the classical era. At

FIG. 5.4 Frontispiece to Joannes Wolters's edition of *Strabo's Geography* (1707). Library of the Holy Monastery of Docheiariou, Mount Athos.

the bottom of the composition, the oval portrait of the French philologist and commentator Isaac Casaubon literally supersedes Strabo's portrait amid burning incense, bones, shells, and other exotica brought to Europe from past voyages of discovery.

On the frontispiece made by Leiden engraver François van Bleyswijck for Nicolas Gueudeville's *Le nouveau theatre du monde, ou La geographie royale* (1713), a complex iconography unfolds on the dock of a busy port, the launching point for the lucrative European maritime trade in the age of blue-water empires. Chronos (balancing a winged hourglass on his bald head) unveils Asia (possibly a reference to the Dutch East India Company's commerce), while Chorographia measures distances on the globe with her compass (fig. 5.5).[38] All look at History, who with Neptune points to Atlas, who patiently endures the weight of the heavenly sphere on his muscular shoulders. At other times, Chronos is conflated with or replaced by other figures from classical mythology. On the frontispieces to Jean van Keulen's nautical atlases, winged putti draw velvety stage curtains on trafficked seas and Poseidon (drawn in the fashion of Chronos) lifts his veil over the constellations of the celestial globe used by sailors for orientation.[39] In other instances, Time and Fame lift white drapes inscribed with the title of the work. For example, on Simon Van Leeuwen's *Illustrious Holland, or a Treatise on the Origin, Progress, Traditions, State and Religion of Old Batavia* (1685), Fame and Time unveil an allegory of Holland telling Clio, the muse of history (sitting on a globe), the stories of Holland's glory, which she diligently jots down in her book.

Such frontispieces grew especially popular in the Netherlands in the first half of the eighteenth century. According to the lawyer, engraver, and writer Romeyn de Hooghe (1645–1708), "the citizens of the United Provinces were ardent collectors of fine and rare things, including splendidly illustrated and sumptuously bound books. Books were to be found in their houses in large numbers. Numbers... which one would expect in the library of a professor, rather than in the library of an ordinary citizen."[40] Besides

Lifting the Veil of Time   [111]

FIG. 5.5 François van Bleyswijck, *Allegory with Atlas, Father Time, and Zeus, Chorography and History* (1713). Rijksmuseum, public domain.

this interest in rare luxury books, the Dutch were fascinated by large illustrated works of a didactic nature, including books of biblical history, collections of portraits of famous people, and, of course, atlases.

Often characterized as "visual encyclopedias" or "museums on paper," these large volumes of engravings and maps formed part of a long-standing antiquarian culture.[41] Seventeenth-century antiquarianism had taken a Baconian bent. It was no mere collectionism, but a practice connected to an increasingly systematic discovery (or unveiling) of the past to illuminate the present. As an anonymous writer stated, the serious antiquarian was meant to love antiquities, "but 'tis only such as draw the Veil from off the Infancy of Time, and uncover the Cradle of the World."[42] Likewise, after Bacon, pioneering archaeologists such as the Danish scholar Ole Worm warned against giving too much credit to particular authors, as time was "author of all authors, and, therefore, of all authority. For truth is rightly named the daughter of time, not of authority."[43] In other words, if the antiquarian was a "seeker of Truth," time was a giant seamless veil he had the moral duty to lift for the benefit of present and future generations.

Eighteenth-century atlases were at once visual collections, beautiful collectibles, and vessels of "diffuse history" for a general audience.[44] The frontispiece to the first volume of Châtelain's *Atlas historique* (1719) captures the encyclopedic nature of these kinds of works, their moralizing character, and their "telescopic" approach to historical time as a vast, seamless fabric holding events together (fig. 5.6).[45] Here Chronos reappears in the traditional fashion of seventeenth-century painting: he flies fast in the sky, carrying an unveiling Truth on his back. The latter exhibits a "Chronological Table of the World Countries" (centered on Greece and the Peloponnese) to winged Fame, who compiles the *Atlas historique* using a Spain-centered globe as her table. Another female figure (perhaps Clio) stands behind the globe and shows Fame a genealogical tree of the many-branched House of Bourbon, whose various claims to the thrones of France and Spain preoccupied much of western Europe during this time. Medallions featuring the Roman emperors hang from the branches of the tree at right; at Fame's feet, a putto braids a crown of laurel amid discarded crowns,

FIG. 5.6 Frontispiece of the first volume of Chatelain's *Atlas historique* (1719). Courtesy of the Barry Lawrence Ruderman Map Collection, David Rumsey Map Center, Stanford Libraries, https://purl.stanford.edu/hd338hb0617.

scepters, and other insignias of regal power. Time features as a mediator between the classical world and modern Europe; between the vicissitudes of history and the broadening horizons of human knowledge; between earth and heaven; between collecting and discovery. Truth and Time fly high over the globe and over the marine horizon as impartial judges of mundane affairs.

First published anonymously in Amsterdam in 1705 (possibly by the Huguenot publisher Zacharias Châtelain), then reprinted in revised editions by his sons, the ambitious seven-volume atlas was among the first to be explicitly called "historical." Its subtitle declared it to be a "new introduction to history, chronology, and ancient and modern geography, set out in new maps in which notice is given of the establishment of the states and empires of the world, their duration, their fall, etc." Despite the name, however, the atlas was scarcely original. As a contemporary reviewer observed, it drew from the major Dutch collections of the previous century and shared their geographical program.[46] Contents were arranged geographically, rather than chronologically, and maps of the contemporary world outnumbered historical ones. In Walter Goffart's words,

> Historical time is more telescoped than ever: the map of the four ancient empires singles out the homelands of the invading peoples held responsible for the fall of the Roman Empire; the Greek empire indiscriminately includes the tracks of the Argonauts, Xenophon's 10,000, Alexander the Great, and Saint Paul; the map of the Roman Empire at its height comes complete with the traces of Julius Caesar's campaign against Pompey as well as comments concerning—again—the dangerous barbarians to the north and south.[47]

More than anything else, the *Atlas historique* was a beautifully presented didactic compendium intended for a general public fascinated by the recently established colonies and new discoveries. Distant lands, such as the Americas, Mongolia, China, and Japan, take a prominent place in the collection. These newly "unveiled" worlds were part of the same universal narrative that began in the biblical and classical worlds and was progressively unfolded by Father Time.[48]

Atlas frontispieces played a key role in this narrative of "unfolding discovery," while at the same time fulfilling other functions. According to the early modern Dutch painter Gerard de Lairesse, the frontispiece served for "the enjoyment of every beholder's eye, the honour and glory of the author as well as the artist and the advantage of the booksellers . . . True, if it is a bad book, the frontispiece will not make it a good one; but the proverb applies here that if a thing is nicely packaged it is already half sold; and therefore it is necessary painstakingly to observe excellence in all things."[49] As physical spaces, frontispieces and title pages, like stage curtains, built up expectation. They stirred imagination. In a sense, the frontispiece was itself a curtain—a paper curtain or veil lifted by an impatient reader curious to discover new worlds of knowledge.

Cartographic Veils

While Mnemosyne looks backward through the veil of oblivion into the past, trumpeting Fame "looks forward to future generations who will preserve forever an event deemed to be unforgettable."[50] Sometimes paired with Time, allegories of Fame lifting or holding veils and drapes were used on frontispieces and title pages to conjure up a sense of futurity and celebration. Within the book, opening stage curtains, flying blankets, and unfurling banners inscribed with page titles renewed discovery and expectation at the turn of each page—like different acts of the same drama trumpeted by Fame.

On the title page of the 1687 German edition of Jean Chardin's *Journal du Voyage du Chevalier Chardin en Perse & aux Indes Orientales*, for example, the heavy drape lifted by Fame and Mercury over distant horizons (fig. 5.7) is echoed in the many flying veils and blankets that populate the following pages of the book.[51] Floating high in the sky, these devices contain the names of the locales represented in the bird's-eye views below, as well as maps and ground plans of their buildings and monuments. The function of these veils is at once informative and rhetorical: appropriately, the "God's-eye" view provided by the map (or ground plan) comes from heaven by way of winged messengers.

The use of cartographic veils intensified in the eighteenth century. They appeared on large-scale maps celebrating military victories, religious power, or simply national or civic pride.[52] On Sébastien de Pontault's *Glorieuse campagne de Monseigneur le duc d'Anguyen commandant les armées de Louis XIII*

FIG. 5.7 German edition of 1687 of Jean Chardin, *Journal du Voyage du Chevalier Chardin en Perse & aux Indes Orientales*. American School of Classical Studies at Athens, Gennadius Library.

(1644), for example, cartographic and bird's-eye views of the battle scene are juxtaposed by way of a sumptuous drape embroidered with the Catholic fleur-de-lis.[53] Territory is literally turned into a French war banner and a war banner into French territory. Such trompe l'oeils allowed mapmakers to bring different perspectives, scales, and temporalities together in the same image; to unfold stories within stories; to craft visual *ekphraseis* reminiscent of Rubens's tapestry series, *The Triumph of the Eucharist* (ca. 1625–1633), which likewise unfolded amid spectacular opening drapes.[54] As with Rubens's sumptuous clothes combining glorification of the Church

Lifting the Veil of Time [115]

with magnification of individual political power, cartographic veils generally fulfilled a celebratory function. The textile medium boosted the sense of triumphant progression and theatricality attached to military campaigns and to the discovery of the past and new worlds.

Military, historical, and geographical celebration combine and peak in the work of the Italian Vincenzo Coronelli, the official cosmographer of the Republic of Venice, and surely one of the most prolific and original experimenters with maps, drapes, and veils of his time. Famous for the two monumental globes he built for Louis XIV and for his vast production of geographical engravings (over 7,000 in total), Coronelli founded the Accademia degli Argonauti (1680), which has been called Europe's first geographical society. It was under its imprint that Coronelli's enormous output of geographical texts, globes, and maps was published.[55] Characteristically, the emblem Coronelli designed for the Accademia features trumpeting Fame unfurling a banner with the motto *Plus Ultra* ("Further Beyond") over the globe, while geographical and navigational instruments decorate the image, which is, in turn, sometimes contained within a giant unfolding drape. Within the image, writes geographer Denis Cosgrove, "a ship surmounts the globe's graticule, while Hercules' club and bearskin cast below suggest that the Moderns have superseded the greatest of the classical heroes, passing beyond the Pillars named for him and reaching to the ends of the earth."[56]

This rhetoric of progressive global discovery permeates the pages of Coronelli's monumental *Atlante veneto* (1690–1701), a thirteen-volume work that was meant as the continuation of the seventeenth-century Dutch cartographer Joan Blaeu's major cartographic publication, *Atlas maior*. Bringing together ancient and contemporary geographical information, Coronelli conceived his magnum opus as "a geographical, historical, sacred, profane, and political description of the empires, kingdoms, provinces and states of the universe, their divisions and borders, with the addition of all newly discovered lands, and expanded by many newly published geographical maps."[57] The *Atlante* shares in the cosmographic dream of total knowledge and presents a spatialized approach to time and universal history; in Cosgrove's words, it reveals cosmographic knowledge "as a hierarchy of representation, proceeding from the divine *Fiat*, through 'the order of Creation' to the scale of individual regions and cities (*iconografia*), palaces (*scenografia*), and rivers (*potomografia*).[58] Different scales are often juxtaposed by means of cartographic veils and similar devices, while flying banners, unrolling scrolls, and undulating drapes simultaneously reveal old topographies and new discoveries.

On the map of Africa (fig. 5.8), for example, Fame flies over the mysterious sources of the Nile as they are being unveiled under a large cloth inscribed with ancient and contemporary speculations about its sources. The scene is balanced by another prominent veil forming the cartouche on the facing page, revealing and yet half-concealing the continent's exotic plant and animal species.

Veils also obscure the past. At the outset of his description of the city of Ravenna, for example,

FIG. 5.8 Detail from the map of Africa in Vincenzo Coronelli's *Atlante veneto* (1690–1701). Courtesy of the David Rumsey Map Collection, David Rumsey Map Center, Stanford Libraries, https://purl.stanford.edu/gv420dv9515.

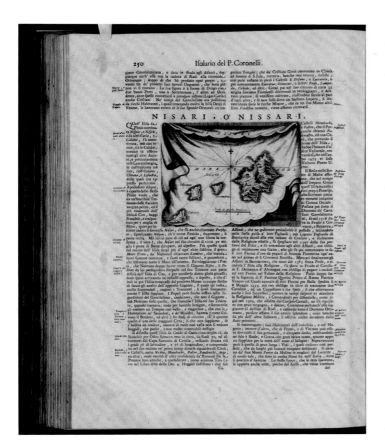

FIG. 5.9 Maps of Nisari (near Rhodes) and Lango (Kos) in Coronelli's *Isolario dell'Atlante veneto* (1696). Biblioteca Nazionale Marciana, Venice.

Coronelli observes how "the memories of the foundation of the city are buried under the darkness of a decrepit antiquity, to such an extent that instead of shedding light on the origin through the whiteness of their sheets, authors further adumbrated it with the darkness of their ink."[59] It is the cosmographer's task to penetrate this dark layer of misleading ink and bring truth to light: in the accompanying illustration, a drape reveals a plan of the ancient city in utmost clarity, with each of its religious buildings carefully mapped and numbered. Similar metaphors proliferate in the descriptions of the Greek islands, whose glorious memories are likewise "buried" under the veil of oblivion, due to the neglect of their current inhabitants oppressed by the Ottoman yoke.[60] Once home to the Knights Hospitaller but now reduced to ruin, the islands near Rhodes (fig. 5.9) are charac-teristically portrayed on thin veils ready to fly away from the page, as if they were ephemeral memories of themselves and their glorious past.

Coronelli's cartographic production and theatrical use of cartographic veils acquired further momentum in the context of the Ottoman-Venetian War of the Morea (1684–1699). The conflict, which saw the creation of a multinational Holy League under the auspices of the Vatican and the Venetian recapture of the Peloponnese and some of the adjacent islands, stimulated an unprecedented flourishing of propagandistic narratives inspired by classical mythology, of which Coronelli was at the forefront. During the war, his workshop restlessly produced hundreds of maps and views of forts and territories seized by the Venetians. In his *Memorie istoriografiche* (1685) and *Teatro della Guerra* (n.d.), many of whose plates were recycled in the *Atlante*'s *Isolario veneto*, the Morea appears as a cartographic stage always observed from a distance and mastered by allegories of Venice and by the viewer from a high-oblique angle (fig. 5.10). Battles and newly conquered lands rapidly unfurl at the turn of each page, as a succession of thin temporal veils. Cartographic veils serve to unfold territorial conquest and the progressive expansion of Venetian domains.

Cartographic veils also act as theatrical "curtains" revealing distant lands to the inhabitants of the motherland (fig. 5.11).[61] The striking visual tactility of these trompe l'oeil illusions has the effect of displacing and intriguing viewers, thus increasing the aesthetic appeal of the atlas. It invites the beholder to touch the ripples and the folds of the cartograph-

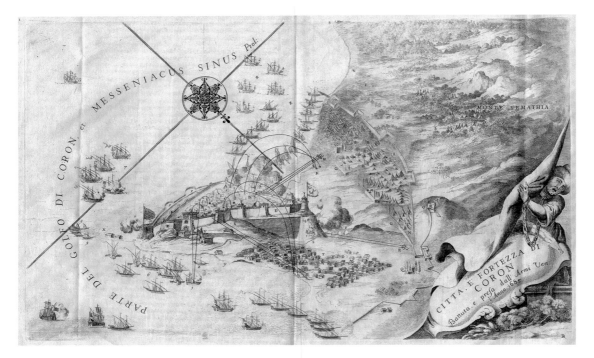

FIG. 5.10 City and fortress of Coron, in Coronelli's *Memorie Istoriografiche delli Regni della Morea, e Negroponte e luoghi adiacenti* (1686). Biblioteca Nazionale Marciana, Venice.

FIG. 5.11 Regno di Negroponte, in Coronelli's *Isolario dell'Atlante veneto* (1696). Biblioteca Nazionale Marciana, Venice.

FIG. 5.12 Portrait of Vincenzo Coronelli on the frontispiece of his *Atlante veneto* (1690–1701). Biblioteca Nazionale Marciana, Venice.

ic veils, even while prioritizing vision.⁶² At the same time, these illusory fabrics signpost the visual narrative, while adding further dynamism to it; or rather, time is represented as fabrics—as a constant process of unveiling triggered by the movement of the reader's hand.

Characteristically, on his terrestrial globe Coronelli portrayed himself in the process of being unveiled by two putti, just as his maps were. On the frontispiece to the *Atlante*, the cosmographer's unveiled figure acts itself as a veil (fig. 5.12). He gazes benevolently at us, his right hand covering the North Pacific, preventing us from seeing what he is pointing to with his compass. The cosmographer's hand fulfils the function that Time's veil did in Wolters's 1707 edition of Strabo (fig. 5.4). Its protective gesture marks a liminal moment in Western intellectual history: the close of the great Age of Discovery and of cosmography. From now on, as the Dutch frontispieces remind us, geographical discoveries and the secrets of the planet shall no longer be unveiled by the omniscient cosmographer, but by Father Time alone.

Conclusions

Wrapping Mnemosyne and naked Truth since classical antiquity, Time's veil fulfilled multiple functions in Baroque Europe. It helped build up expectation; signpost narratives of progress; and give visual and metaphorical expression to a telescoped history in

which different eras coexisted in the same space, as threads of the same seamless fabric. The veil was therefore a paradoxical device in that it conjured up progression and simultaneity; it at once revealed and concealed. Morphed into map and positioned in the interior of the book, the veil was not simply a compelling metaphor rendered visually, but a powerful material tool for *producing* time by way of spatialized performance. Whether following the unfolding of a journey or a war, of geographical discovery or universal history, time was enacted as the reader leafed through the pages, sequentially lifting cartographic veils. Time's veil enlisted a collaboration among the eye, the hand, and the physical medium.

By the nineteenth century, as a nearly fully mapped globe was starting to shrink, Time's veil abandoned atlas frontispieces and maps. Having lost its spatial qualities, it morphed into an ambiguous and mutable entity. In a speech delivered in occasion of the first transatlantic telegraph communication in 1858, American colonel John S. Preston characteristically declared that "Space has been conquered, and voices swifter than chariot-wheels of Phoebus are sent beneath the seas and around the Earth: but the Veil of Time within its mysterious folds still holds and hides the vast results."[63] The laying of the transatlantic cable announced a new global geographical imagination in which space had been tamed, for communications across the Atlantic now traveled instantaneously. Time nonetheless remained an ungraspable entity: its folds were mysterious, obscure, silent.[64] Yet only ten years later, in 1868, the Society of Antiquaries of London selected Time's veil as the keystone for the entrance of its new building in Piccadilly. No longer a heavy drape awaiting to be lifted by a muscular Father Time, the veil was now a diaphanous, almost ethereal presence revealing the delicate traits of Mnemosyne's face—as though, in the age of professional archaeology and geology, the veil shrouding the mysteries of the past had now become transparent or, in Schopenhauer's words, "the thinnest of veils."[65]

Notes

I would like to thank Jeremy Brown and the organizers and participants of the Time in Space conference for their precious input and suggestions on previous drafts of this chapter.

1. Walter Goffart, *Historical Atlases: The First Three Hundred Years, 1570–1870* (Chicago: University of Chicago Press, 2003), 38. To historical maps one should also add "time maps," as discussed in Daniel Rosenberg and Anthony Grafton, *Cartographies of Time: A History of the Timeline* (New York: Princeton Architectural Press, 2010) and, more broadly, allegorical maps embedding a temporal dimension, such as eighteenth-century French and British "maps of life" and "matrimonial maps." See Franz Reitinger, "Mapping Relationships: Allegory, Gender and the Cartographical Image in Eighteenth-Century France and England," *Imago Mundi* 51 (1999): 106–30; and Jeffrey Peters, *Mapping Discord: Allegorical Cartography in Early Modern French Writing* (Newark: Delaware University Press, 2004).

2. Christian Jacob, *The Sovereign Map: Theoretical Approaches in Cartography Throughout History* (Chicago: University of Chicago Press, 2006), 326.

3. Basil of Caesarea, *Hexaemeron*, 6.1.

4. Mark Johnson, *The Body in the Mind: The Bodily Basis of Meaning, Imagination, and Reason* (Chicago: University of Chicago Press, 1987).

5. "Velut magnum Carmen cuiusdam ineffabilis modulatoris" (*Epist.* 138.5). The passage is discussed in Pierre Hadot, *The*

Veil of Isis: An Essay on the History of the Idea of Nature (Cambridge, MA: Harvard University Press, 2006), 201.

6. Charles Dickens, *Hard Times for These Times* (New York: Hurd and Houghton, 1869), 129; Henry David Thoreau, *Walden* (1854; repr. New York: Cosimo Classics, 2009), 64.

7. Joseph Campbell, for example, describes a statue of the Hindu goddess Shiva holding in her hand "a flame which burns away the veil of time and opens our minds to eternity." See Harry Eiss, *Divine Madness* (Newcastle: Cambridge Scholars Publishing, 2011), 345.

8. Likewise, the word "discovery" was used by the Portuguese to designate the finding of new territories during the age of geographical exploration, but became common currency in the English language in the late sixteenth century and first half of the seventeenth century, when it had at least six meanings, including "to uncover," "disclose," or reveal. See Jerry Brotton, *The History of the World in Twelve Maps* (New York: Penguin Books, 2013), 155–56.

9. James Spedding, ed., *Works of Francis Bacon*, 14 vols. (London: Longman, 1861–1879), vol. 8, p. 257.

10. Denis Cosgrove, "Global Illumination and Enlightenment in the Geographies of Vincenzo Coronelli and Athanasius Kircher," in *Geography and Enlightenment*, ed. David Livingstone and Charles Withers (Chicago: University of Chicago Press, 1999), 48.

11. Alain Schnapp, *The Discovery of the Past* (New York: Abrams, 1997).

12. See Paul Hoftijzer, "The Dutch Republic, Centre of the European Book Trade in the 17th Century," EGO [European History Online] (2015), http://ieg-ego.eu/en/threads/backgrounds/the-book-market/paul-g-hoftijzer-the-dutch-republic-centre-of-the-european-book-trade-in-the-17th-century#InsertNoteID_3_marker4.

13. According to Andrew Pettegree and Arthur der Weduwen, "The Dutch book world of the seventeenth century was built on seven main markets: state communication, printing pamphlets and broadsheets for over 105 jurisdictions scattered through the seven provinces; printing for universities and illustrious schools; church books (Bibles and psalters) and devotional literature; poetry and literature; school books; small-format Latin works for the export market; and news. Together, these accounted for well over 90 per cent of the more than 360,000 editions published in the Dutch Republic before the end of the seventeenth century." See *The Bookshop of the World: Making and Trading Books in the Dutch Golden Age* (New Haven, CT: Yale University Press, 2019), 397.

14. See, for example, the Louvre sarcophagus (second century CE) and a statue of the same period retrieved near Tivoli, described in Ennio Quirino Visconti, *Il Museo Pio Clementino* (Milano: Bettoni, 1819), 173–74.

15. Richard Onians, *The Origins of European Thought about the Body, the Mind, the Soul, the World, Time, and Fate* (Cambridge: Cambridge University Press, 1954), 344. See also Penelope's stratagem to deceive her suitors: by weaving and unweaving her shroud, she pauses time (*Odyssey*, book 16).

16. Onians, *The Origins of European Thought*, 389.

17. John Scheid and Jesper Svenbro, *The Craft of Zeus: Myths of Weaving and Fabric*, trans. Carol Volk (Cambridge, MA: Harvard University Press, 2001), 159.

18. *The Cosmographia of Bernardus Silvestris*, trans. Winthrop Wetherbee (New York and London: Columbia University Press, 1973), 74; and Dante Alighieri, *Paradise*, 16.7–9.

19. In the Regensburg drawing of around 1100, Chronos/Saturn wears a big, fluttering veil. See Erwin Panofsky, *Studies in Iconology: Humanistic Themes in the Art of the Renaissance* (1939; repr. San Francisco: Harper and Row, 1972), 75.

20. Cesare Ripa, *Iconologia* (Siena: Heredi di Matteo Florini, 1613 [1593]), 298.

21. Donald Gordon, "'Veritas Filia Temporis': Hadrianus Junius and Geoffrey Whitney," *Journal of the Warburg and Courtauld Institutes* 3 (1940): 228–40; and Soji Iwasaki, "*Veritas Filia Temporis* and Shakespeare," *English Literary Renaissance* 3 (Spring 1973): 249–63, at 250.

22. "[Time,] that shuttle brained tall long man / that nere standeth still but flyghth as fast as he canne / muche like as he swymmed or glided upon yce? / . . . He carried a clocke on his heade / A sandglasse in his hande" (*Respublica*, 1553). The passage is discussed in Gordon, "'Veritas Filia Temporis,'" 229.

23. Erasmus, *Adagia*, 2.4.17, in *The Adages of Erasmus*, trans. William Barker (Toronto: University of Toronto Press, 2001), 179–81.

24. Panofsky, *Studies in Iconology*, 89.

25. Fritz Saxl, "Veritas Filia Temporis," in *Philosophy and History: Essays Presented to Ernst Cassirer*, ed. Raymond Klibansky and H. J. Paton (Oxford: Clarendon Press, 1936), 197–222 (quotation from 209).

26. For example, Robert Greene's romance of 1588 *Pandosto* is subtitled "The Triumph of Time," with the tag "*Temporis filia veritas*" on the title page. This work inspired Shakespeare's *Winter's Tale*, in which a daughter, in time, teaches truth to her father.

27. The popularization and wide spread of the theme across Europe owes to texts such as Andreas Alciatus's *Emblemata* (1542, 1602) and Ripa's *Iconologia* (1593, 1603) and their various editions and translations, including the influential Dutch translation by Dirck Pietersz Pers (1644).

28. Paul Huys Janssen, "Theodor van Thulden: Time Revealing Truth" (http://www.steigrad.com/thulden-time-revealing-truth/). At other times, Truth is conflated with allegories of painting and sculpture. For example, in Henri Testelin's sketch *Time Assisted by the Love of Virtue Dispels the Truth of Painting out of the Clouds of Ignorance* (1655), Time lifts a mantle of clouds over Truth, who, in turn, unveils herself and her palette. Nicolas Loir likewise portrays female allegories of Painting and Sculpture unveiled by Time (1663), a motif he also uses in a painting celebrating the foundation of the French Royal Academy of Painting and Sculpture (ca. 1648). Paolo Dematteis's *Allegory of Knowledge and the Arts in Naples* (dating to the second half of the seventeenth century) embeds a painting of Time unveiling Truth within the painting.

29. Saxl, "Veritas Filia Temporis," 216–18.

30. Her right foot is set on a globe. Intriguingly, between her big toe and England, a thorn reminds the attentive observer of the suffering the expansion of Anglicanism caused the Pope.

31. Panofsky, *Studies in Iconology*, 92.

32. Alexandre Koyré, *From the Closed World to the Infinite Universe* (Baltimore: Johns Hopkins University Press, 1957), 2.

33. On the frontispiece of the Dutch naturalist and microscopist Anton van Leeuwenhoek's *Anatomia seu interiora rerum* (1687), the unveiling of monstrous Isis is facilitated by Time (the old man on the right) and the lens of the microscope held by Inquisitiveness. Interesting parallels can be drawn with human anatomical tables, sometimes featuring self-dissecting female bodies as allegories of Truth. See Jonathan Sawday, *The Body Emblazoned: Dissection and the Human Body in Renaissance Culture* (London and New York: Routledge, 1996); and Raphaël Cuir, *The Development of the Study of Anatomy from the Renaissance to Cartesianism: Da Carpi, Vesalius, Estienne Bidloo* (Lewiston, NY: Edwin Mellen Press, 2009).

34. Hadot, *The Veil of Isis*, 129.

35. Basil Montagu, *The Works of Francis Bacon, Lord Chancellor of England*, 3 vols. (Philadelphia: Carey and Hart, 1844), vol. 1, p. 358. Fittingly, the frontispiece of Bacon's *Of the Advancement and Proficience of Learning, or the Partitions of Sciences* (1640) features in a drape hanging over a vessel set against an open horizon.

36. As opposed to classical authors such as Seneca, who used the formula with an exclusive moral meaning ("in order to master anger, once must not react right away, but must give oneself time, for time unveils truth"), or sixteenth-century religious and political appropriations (Hadot, *The Veil of Isis*, 179).

37. Hadot, *The Veil of Isis*, 176.

38. Nicholas Gueudeville, *Le nouveau theatre du monde, ou La geographie royale* (Leiden: Chez Pierre van der Aa, 1713).

39. *Le grand atlas de la mer ou monde Atlantique* (Amsterdam, 1680). The same motif is found in Francesco Primaticcio's *Astronomia* (1552), whereby a star-crowned allegory of Astronomy takes the place of Time and the heavenly sphere that of the earthly globe.

40. Jean C. Streng, "The Leiden Engraver Frans van Bleyswyck (1671–1746)," *Quaerendo* 20, no. 2 (1990): 111–36 (quotation from 114).

41. The Antwerp cartographer Abraham Ortelius, author of the first printed atlas (1570), was himself a collector. Large seventeenth-century atlases were part of private collections and also used as actual portfolios. For example, one Amsterdam

lawyer named Laurens van der Hem inserted between the pages of his copy of Blaeu's *Atlas maior* (1662) more than 1,800 city views, seascapes, architectural sketches, ethnographic prints and portraits, and so on. See Diane Dillon, "Consuming Maps," in *Maps: Finding Our Place in the World*, ed. James R. Akerman and Robert W. Karrow Jr. (Chicago: University of Chicago Press, 2007), 296.

42. Schnapp, *Discovery of the Past*, 233.

43. Schnapp, *Discovery of the Past*, 164.

44. Goffart, *Historical Atlases*, 22.

45. Châtelain, *Atlas historique, ou nouvelle introduction a l'histoire, à la chronologie & à la géographie ancienne & moderne: Représentée dans de nouvelles cartes, où l'on remarque l'établissement des etats & empires du monde, leur durée, leur chûte, & leurs differens gouvernemens. . .*, 7 vols. (Amsterdam: Châtelain, 1705–1720). See Goffart, *Historical Atlases*, 132, for the complicated problems of authorship for this work.

46. Goffart, *Historical Atlases*, 132.

47. Goffart, *Historical Atlases*, 133.

48. "The equivalence of 'general' or 'universal' atlases with 'historical' ones remained standard into the nineteenth century, so much so that historical atlas in the new sense it was acquiring—a collection of specialized maps for history—sometimes clashed with general atlases and their familiar, diffuse custom" (Goffart, *Historical Atlases*, 23).

49. Quoted in Streng, "The Leiden Engraver," 115.

50. Aleida Assmann, *Cultural Memory and Western Civilization: Functions, Media, Archives* (Cambridge: Cambridge University Press, 2011), 39.

51. Jean Chardin, *Des vortrefflichen Ritters Chardin* (Leipzig: Gleditsch, 1687).

52. An interesting example is found on the frontispiece of Athanasius Kircher's *China Illustrata* (1667). Later examples include the view of Prague (also portrayed on a drape) embedded in Johann Christoph Müller's lavish twenty-five-sheet map of Bohemia (1720) and various urban maps, such as Antonio Espinosa's nine-sheet topographical plan of Madrid (1769) and Giovanni Battista Nolli's famous map of Rome produced in collaboration with Piranesi (1748). In these urban maps, the cartographic veil features as a sort of "skin" peeled away from the glorious ruins of ancient and mythical pasts. These maps are discussed in Peter Barber and Tom Harper, *Magnificent Maps: Power, Propaganda, and Art* (London: British Library, 2010), 70–73; and Brigitte Marin, "Il Plano Topografico di Madrid di Antonio Espinosa de los Montoros (1769): monarchia riformatrice, nuovi saperi della città e produzione cartografica," in *Le città dei cartografi: Studi e ricerche di storia urbana*, ed. Cesare De Seta and Brigitte Marin (Naples: Electa Napoli, 2008), 148–68.

53. This map is discussed in Alberto Dragone, *Segni e sogni della terra: Il disegno del mondo dal mito di Atlante alla geografia delle reti* (Milan: De Agostini, 2001), 127.

54. For the date of the tapestry series, see Anne T. Woollett, "Faith and Glory: The Infanta Isabel Eugenia and the *Triumph of the Eucharist*," in *Spectacular Rubens: The Triumph of the Eucharist* (Los Angeles: Getty Publications, 2014), 11.

55. Denis Cosgrove, "Global Illumination and Enlightenment in the Geographies of Vincenzo Coronelli and Athanasius Kircher," in David N. Livingstone and Charles W. J. Withers, eds., *Enlightenment Geographies* (Chicago: University of Chicago Press, 2000), 37.

56. Cosgrove, "Global Illumination," 41.

57. Vincenzo Coronelli, *Atlante veneto*, vol. 1 (Venice: Domenico Padovani, 1690), title page; English translation from Cosgrove, "Global Illumination," 39.

58. Cosgrove, "Global Illumination," 39.

59. Vincenzo Coronelli, *Isolario, descrittione geografico-historica, sacro-profana, antico-moderna, politica, naturale, e poetica. Mari, golfi, seni, piagge, porti, barche, pesche, promontorj, monti, boschi, fiumi . . . ed ogni piu esatta notitia di tutte l'isole coll'osservationi degli scogli sirti, scagni, e secche del globo terracqueo. Aggiuntivi anche i ritratti de' dominatori di esse. Ornato di trecento-dieci tavole geografiche, topografiche, corografiche, iconografiche, scenografiche a' maggiore dilucidatione, ed uso della navigatione, et in supplimento dei 14 volumi del Bleau. Tomo 2. dell'Atlante veneto*, pt. 1 (Venice: A' Spese Dell'Autore, 1696), 69.

60. Coronelli, *Isolario*, 163.

61. Veronica della Dora, "Mapping 'Melancholy-Pleasing Remains': Morea as a Renaissance Memory Theater," in Sharon

E. J. Gerstel, ed., *Viewing the Morea: Land and People in the Late Medieval Peloponnese* (Washington, DC: Dumbarton Oaks Research Library and Collection, 2013), 455–75. Intriguingly, transposed on fabrics, islands and cities become trophies akin to war banners, like the Ottoman one the Venetians captured at Coroni (to which Coronelli dedicated two separate engravings). The orientation of the maps in these works reflects their strategic importance. For example, Negroponte and Crete are portrayed horizontally, occupying two full pages.

62. David Weimer, "Look but Don't Touch: Eighteenth-Century Cartographic Illusions of Tactility," paper presented at the International Conference on the History of Cartography, Belo Horizonte, July 9–14, 2017.

63. *Report of the Dinner Given by the Americans in Paris, August the 17th, at the "Trois Frères," to Professor S. F. B. Morse, in Honor of His Invention of the Telegraph, and on the Occasion of Its Completion under the Atlantic Ocean* (Paris: E. Brière, 1858), 16.

64. In 1847, the French archaeologist Boucher de Perthes made similar use of the metaphor: "On this path of discovery we are only at the point of departure. So why say that we have reached the end of the voyage? Because we have lifted a corner of the veil, must we conclude that we have seen all that the veil conceals?" (quoted in Schnapp, *The Discovery of the Past*, 372). See also John Brady: "Wrapt in the veil of Time's unbroken gloom, Obscure as death, and silent as the tomb, Where oblivion holds her dusky reign, Frowns the dark pile on Sarum's lonely plain. Yet think not here with classic eye to trace Corinthian beauty": see "Stonehenge," in John Brady, *Varieties of Literature* (London: Geo. B. Whittaker, 1826), 237.

65. Robert Wicks, *Schopenhauer* (Oxford: Blackwell, 2008), 77.

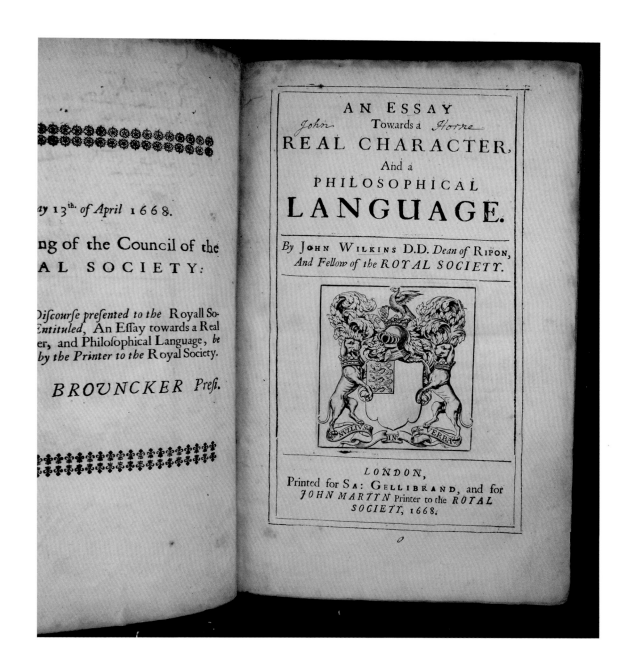

FIG. 6.1 John Wilkins, title page, from *An Essay Towards a Real Character* (1668). Copy previously owned by John Horne Tooke and C. K. Ogden. The William Andrews Clark Memorial Library, University of California, Los Angeles.

6

Daniel Rosenberg

A Map of Language

The notion that languages may be viewed as conceptual maps was of particular importance in the seventeenth and eighteenth centuries in Europe, when the construction of good taxonomies was understood as paramount to philosophy and natural philosophy alike. In this period, Michel Foucault famously argued, words themselves formed "a table on which knowledge [was] displayed in a system contemporary with itself."[1] Among the most notable of the actual word tables of the seventeenth century was the one at the heart of the "real character" system proposed in 1668 by John Wilkins, polymath founder of the Royal Society of London.[2] This artificial language system, Wilkins argued in his *Essay Towards a Real Character*, offered important advantages over "instituted languages" such as Latin, French, or English (fig. 6.1).[3] In comparison with all of these, Wilkins's system was compact, regular, and rational. As such, it would not only be easy to learn and to use, it would also provide a tool for clarifying relationships among things and ideas and for distinguishing between good reasoning and bad. At the heart of the system was a classified table of root terms, from which, in principle, could be constructed expressions for every idea and thing in the universe, both extant and possible (fig. 6.2).[4]

From the beginning, Wilkins's table was a focus in assessments of his system. For example, in his 1800 book *Des signes et de l'art de penser*, the French language theorist Joseph-Marie Degérando argued that the conceptual arrangement of terms that underlay the table was arbitrary. Wilkins ought to have been classifying the world; instead, he was simply dividing it up. In Degérando's view, Wilkins was acting as a *cartographer* of language rather than as a *natural philosopher*. He writes:

> Division differs from classification in that the latter bases itself upon the intimate properties of the objects it wishes to distribute, while the former follows a rule to a certain end to which these objects are destined. Classification apportions ideas into genera, species, and families; division allocates them into regions of greater or lesser extent. Classification is the method of botanists; division is the method by which geography is taught.[5]

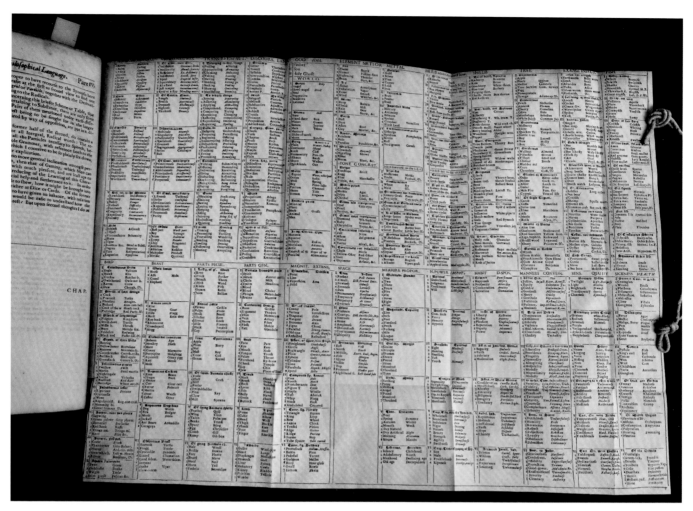

FIG. 6.2 John Wilkins, brief table containing the radicals, from *An Essay Towards a Real Character* (1668). The William Andrews Clark Memorial Library, University of California, Los Angeles.

Degérando meant this as a criticism. Yet his account of Wilkins can also be read descriptively, and as the basis of a different kind of question: if Wilkins was practicing cartography, what kind of cartography was it?

Part of the answer, we know: *pace* Degérando, in the table of terms folded into his *Essay*, Wilkins was attempting to foliate the branches of a conceptual tree laid out in his main text. That "general scheme" of classification was very much the sort of hierarchical framework that Degérando preferred, but, as Degérando accurately reports, when Wilkins lays out his actual lexicon, the results turn out to be messier than his initial scheme suggests (fig. 6.3). This did not disturb Wilkins, for whom the system was a draft, the principal purpose of which was not to name everything in the universe but to lay out a set of basic terms from which, like tinker toys, every other word one needed might be assembled.

This approach introduced play into the system. On the one hand, Wilkins's choices were reasoned, encyclopedic, and guided by a hierarchical "gener-

[128]   CHAPTER SIX

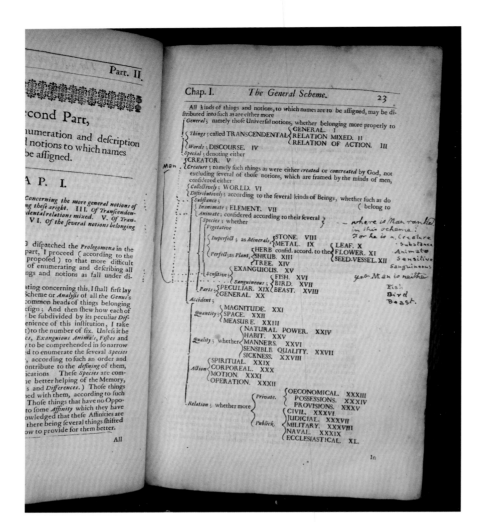

FIG. 6.3 John Wilkins, general scheme of radicals with manuscript notes by John Horne Tooke pointing out the difficulty of locating the entity "man" in the taxonomic framework, from *An Essay Towards a Real Character* (1668). The William Andrews Clark Memorial Library, University of California, Los Angeles.

al scheme."[6] On the other hand, the adequacy of the system turned out to rely less on the philosophical perfection of the underlying taxonomy of roots than it did on their potential to be combined and recombined, like the ciphers that Wilkins discussed in his earlier treatise *Mercury, the Secret and Swift Messenger* (1641).[7] It mattered less *what* these terms were than *how* they could be used. Put another way, these basic words were designed as much to express *relationships* (framed by Wilkins as, for example, *differences*, *opposites*, *affinities*, and so forth) as they were to name *things*. And, to the extent to which they were designed to be built up into names, those names, too, were naturally expressions of relation. This presented opportunities for aspiring taxonomists, which were taken up with energy by Wilkins's collaborators, natural philosophers Francis Willughby and John Ray; it also presented nearly endless choices.[8]

This is an aspect of Wilkins's system that Jorge Luis Borges, in his 1942 essay "The Analytical Language of John Wilkins," already points out.[9] Having promised a radical reduction of language, Wilkins's treatise explodes with terms to be rendered in the new language, ranging from Venus Looking-Glass, Bristol Nonesuch, and Yellow Loose Strife to Goat-Chafer, Death Watch, Lady-Cow, Barble, Gudgeon,

A Map of Language [129]

and Bansticle.[10] To cut through the babble, one might imagine any number of strategies, and during subsequent centuries, a variety of new naming conventions were designed to precisely this end. Some of these, such as the scientific nomenclatures of Antoine Lavoisier and of Carl Linnaeus, met with great success.[11] Wilkins's own system sat upon a basic armature of classification descending from *predicament* to *genus* to *difference* to *species*. Each level of classification below the *predicament*, which in any case was an informal grouping of *genera*, was assigned a list of sounds. Each of the forty *genera* got one, represented in English by a combination of consonant and vowel. Closely related *genera* (often belonging to a common *predicament*) shared initial consonants.[12] The same pattern followed for *differences* and *species*.

Wilkins's main contribution, however, is not his taxonomy as such, though he made an honest effort to produce something useful in this regard; rather, it is the articulation of a set of principles by which any possible system of nomenclature would reveal its own logic and quality. Wilkins writes: "And tho' it should be of no other use but this, yet were it in these days well worth a man's pains and study, considering the Common mischief that is done, and the many impostures and cheats that are put upon men, under the disguise of affected insignificant Phrases."[13] The special virtue of Wilkins's system, then, is its ability to reveal the ontological schemes hidden in language, and, in particular, in those "pretended, mysterious, profound notions, expressed in great swelling words."[14] This happens spontaneously in the articulation of the taxonomy, with all of its inconsistencies, redundancies, and curiosities.

This is, of course, what attracted Borges to Wilkins. It is in his essay on Wilkins that Borges spins the famous tale of the "Chinese encyclopedia." And, by the same token, this is what attracted Foucault, whose *Order of Things* reiterates Borges's story of the Chinese encyclopedia, to Borges.[15] Borges writes:

> I have noted the arbitrariness of Wilkins . . . obviously there is no classification of the universe that is not arbitrary and conjectural. The reason is very simple: we do not know what the universe is. . . . We must go even further; we must suspect that there is no universe in the organic, unifying sense inherent in that ambitious word. If there is, we must conjecture its purpose; we must conjecture the words, the definitions, the etymologies, the synonymies of God's secret dictionary. But the impossibility of penetrating the divine scheme of the universe cannot dissuade us from outlining human schemes, even though we are aware that they are provisional.[16]

The closer one examines Wilkins's vocabulary, the more its organizational logic frays, and the more categories pile up on one another. If one wished to fashion a conceptual map for seventeenth-century England on the basis of Wilkins's word lists, there is certainly plenty of material. Just to take an example, the part of his word tree devoted to state-sanctioned forms of murder—*beheading, quartering, dissecting, stoning, shooting, pressing, precipitating, stabbing, em-*

paling, starving, paining, stifling, burying alive, drowning, burning alive, crucifying, and *breaking on the wheel*—suggests a highly nuanced and codified cultural framework.[17] Still, while his "general scheme" is impressive in extent, the closer one gets to individual terms, the less the system acts like an analytic, and the more it acts like, well, a language.

Remarkably, in his own hierarchical scheme, Wilkins finds no rationally satisfying home for the terms *map* or *diagram*. In fact, he places both in a "very Heterogenous heap" that he calls *common mixed materials*, including also *provisions for cattle*, *treatments for wounds*, and *tools of writing*.[18] Within this heap, *map* and *diagram* are grouped with other sorts of *visual ornament*. Like the styles of execution, these are evocative of the times. Wilkins includes among them *picture, portrait, effigy, draught, chart, landscape, image, projection, scheme, analem, arras, enammel, image, statue, puppet, idol, coloss,* and *crucifix*.[19] Wilkins recognizes—and his system embodies—the fact that every system of signs must be in some ways arbitrary. But consider the alternative: "instituted languages" are almost entirely so.[20]

Thus, while there are some surprising elements in Wilkins's system—none so fantastic as Borges imagined or as Degérando feared—in context, the work is striking for its pragmatism. The system is, in Wilkins's terms, "philosophical," because it operates upon the universe by a formal analogy, not because it contains it in microcosm. Wilkins employs structures in language such as opposition and affinity to code the same kinds of relationships in the things that his lexicon represents.[21] This is also to say that if we wish to think of Wilkins's project as a kind of linguistic cartography, we must think of it as a cartography of a particular, constructive sort—as a tool for articulating relationships, for putting things in places, and for keeping these places organized and addressable. We might think of it as a *program* for language, in both senses of the term.

It is this characteristic of Wilkins's essay that the modern critic Umberto Eco finds visionary. In Wilkins's system, Eco notes, it is not only plausible but necessary for complex things and ideas to appear in multiple places in the system of organization. This does not make the system wrong. It makes it *useful*. And, in view of more recent developments in information design, it makes it forward-looking, as well. Eco asks:

> What if we regarded the defect in Wilkins' system as its prophetic virtue? What if we treated Wilkins as if he were obscurely groping towards a notion for which we have only recently invented a name—hypertext? A hypertext is a program for computers in which every node or element of the repertory is tied, through a series of internal references, to numerous other nodes. . . . If this were the case, many of the system's contradictions would disappear, and Wilkins could be considered as a pioneer in the idea of a flexible and multiple organization of complex data, which will be developed in the following century and in those after.[22]

Whether or not one agrees fully, Eco's impulse resonates with an important tradition of reading Wilkins

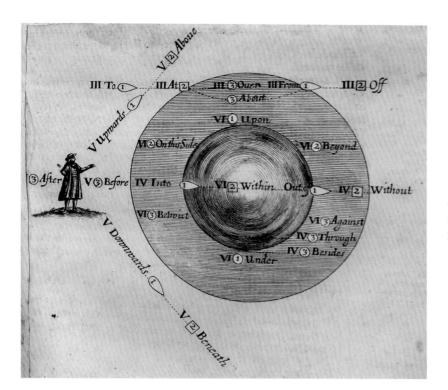

FIG. 6.4 John Wilkins, diagram of local prepositions, detail, from *An Essay Towards a Real Character* (1668). The William Andrews Clark Memorial Library, University of California, Los Angeles.

that runs from John Horne Tooke in the eighteenth century to C. K. Ogden in the twentieth, who, as Lydia Liu has argued, understood Wilkins's approach as combinatorial in a way related to the operation of symbolic machines.[23]

What if, following this train of thought, we regard Wilkins's system not as a static array of locations but as a glimpse of possible epistemological paths or as a potential repertoire of moves?[24] Need we abandon the analogy between his linguistic theory and cartography? Or might we imagine a different framework of mapping, perhaps more choreographic than chorographic, to which the full range of his ideas belongs? To test the virtues of continuing to see Wilkins's system as a kind of cartography, it may be useful to look in more detail at the visual strategies of the *Essay Towards a Real Character* and in particular at the figures in the volume. They are not many, but they are beautiful and telling. Most are tables, including the scheme of radicals. Others present the written

characters that Wilkins invented to go along with his system. These are the *real characters* referred to in the title of the book. Each corresponds to a root word in the lexicon. Another shows the different positions that the human vocal organs take when producing sounds.

One image differs greatly from all of these: Wilkins's diagram of what he calls the *local prepositions* (fig. 6.4).[25] Notably, it is Wilkins's only attempt to directly diagram a part of speech or a function of language, and for this reason, it is of special importance here. For this reason, too, it attracted the attention of many later language theorists. It was copied almost exactly into the great linguistic opus of the English eighteenth century, John Horne Tooke's *Epea Pteroenta; or The Diversions of Purley* (2 vols., 1786, 1805), and it was the only diagram in that work.[26] And it formed the basis for many later diagrams, up to and including the diagrams of prepositions in C. K. Ogden's and I. A. Richards's books on

Basic English, in the 1930s.[27] Going even further than Wilkins in their reduction of the lexicon, Ogden and Richards distilled the vocabulary of English to only 850 words. Ogden, who was an aficionado of Horne Tooke, as well as of Wilkins, owned Horne Tooke's annotated copy of Wilkins's *Essay*, from which, Ogden remarked in his own marginal notes, the diagram of the local prepositions had been excised in order to be copied into Horne Tooke's own book (fig. 6.5).

Wilkins's diagram of prepositions is curious in many ways. We have previously considered his grand table of words as a kind of conceptual map. Here, we are faced with something quite different, a graphic image of something that looks like a space that humans might inhabit. In the diagram, at left, we see a man dressed like a seventeenth-century parson. Perhaps he is meant to be Wilkins himself, who, in addition to his role as secretary of the Royal Society, was an Anglican clergyman and theologian, serving as Bishop of Chester from 1668, the year the *Essay* was published, until his death in 1672 at age fifty-eight. The little man in the diagram stands on a patch of grass. His right arm is crooked at his hip. His left arm gestures forward and a bit up in the direction of two great concentric spheres seemingly floating in space.

Around the little man, across and above the spheres, are twenty-four words. Here, we have something that resembles a map in ways different from Wilkins's table of root words. The diagram presents twenty-four English prepositions—*within*, *betwixt*, *beyond*, and so forth—mapped in a conceptual scheme in relation to the little man and to one another. Not every preposition from Wilkins's system

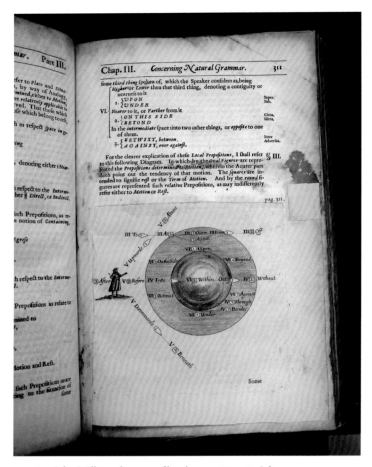

FIG. 6.5 John Wilkins, diagram of local prepositions in John Horne Tooke's copy of *An Essay Towards a Real Character* (1668). The William Andrews Clark Memorial Library, University of California, Los Angeles.

is here. Among his radical words, Wilkins includes thirty-six prepositions. Only those most closely related to questions of space appear in the graphic. In his commentary on the diagram a century later, Horne Tooke would make the case that Wilkins might well have put all thirty-six of his prepositions into the diagram, as ultimately all devolve from spatial ideas.[28]

The same notion was put to the test in C. K. Ogden's explanation of Basic English, which recharacterized all the prepositions as *directives* (signs of direction) and reduced their number further, to twenty-one (fig. 6.6). In his own image and in the

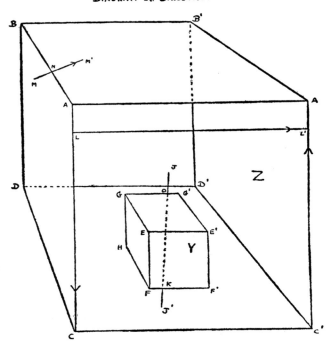

FIG. 6.6 C. K. Ogden, diagram of directives, from *Basic English: A General Introduction* (1930). The William Andrews Clark Memorial Library, University of California, Los Angeles.

accompanying text, Wilkins too suggests that the symbolization of time in prepositions may be only a special case of the symbolization of space.[29] We see this at the left of the diagram, where the word *after* is positioned behind the parson and the word *before* is positioned in front of him. In Wilkins's new language, as in common English, these terms—*before* and *after*—could be used indifferently to indicate temporal and spatial positions.[30]

One implication of this, of course, is that time, like space, ought to be representable in a kind of cartography for which this very image might serve as a key. In the century following Wilkins, as Anthony Grafton and I have argued elsewhere, this intuition was widely developed by historians, astronomers, meteorologists, economists, and other scholars whose work benefited from graphic systems for representing time.[31] During the same period, the issue was addressed, too, by a number of grammarians, resulting in interesting visualizations of time and tense.[32] The notion that a time-map—a chronological timeline, for example—should be thought of as having an underlying visual grammar, and the notion that the flow of language might be visualized as a kind of linear diagram, developed mostly during the eighteenth and nineteenth centuries. In the latter, there are frequent echoes of Wilkins.[33] In relation to all of this, Wilkins's diagram is historically early, and in his period there is not much that really looks like it.[34] In the diagram, Wilkins is evidently working out not only how to diagram language but also how to imagine together in such a diagram the dimensions of space and time.

In his diagram, the demeanor of Wilkins's little parson is staid. He seems to be pointing, but he is not pointing *at* anything. He is drawn in relationship to the words and shapes around him, but he appears oblivious to them. If he were able to see the two floating orbs before him, what would he think? The keys to the diagram are the words which, by their position—as in a calligram—*show* what they *say*.[35] And these words are kinetic. They dance in the diagram: *upwards* and *downwards*, *within* and *out of*, *over*, *about*, and *off*. Most of the words play about within the horizontal band defined by the circles on

the right. *Above* and *beneath* manage an escape to the places they name. Each preposition in the diagram is adorned by a jewel-like shape: a circle, square, or teardrop. Words next to teardrop figures are prepositions of motion; those beside squares are prepositions of rest. Those beside circles refer sometimes to rest and sometimes to motion.

In the center cell, the words *within* and *out of* are trapped in a kind of mutual gravitational lock. The latter—marked by a teardrop—is, despite it all, still within the center sphere. Other terms circulate about. *Upon* is on the center sphere; *under* is beneath it. At upper right, *on this side*; at upper left, *beyond*. *Into*, with its own teardrop, is trying to get where *out of* already is. *Betwixt* opposes *against*. *Through* sits right above *besides*. *To, at, over, about, from,* and *off* skip through the top portion of the circle, like stones across water. The spheres account for three of the categories of prepositions that Wilkins enumerates in his "general scheme": those of "space in general" (III); the "relation of containing" (IV); and things "spoken of" (VI). The prepositions that pertain to the parson relate to what Wilkins calls the "imaginary part of a thing" (V). These are *upwards* and *downwards*, directions that someone or something might take, and *above* and *beneath*, where they might then end up. Finally, fundamentally, the temporal prepositions, *after* and *before*, waggle at the parson's waist like a hula hoop. Without these, he would have a lot of trouble heading *upwards, downwards,* or anywhere else.

Did Wilkins need the little man in his diagram? He was hard to do without. Yet the trend over time was away from him. In the nineteenth century, one grammarian who imitated Wilkins replaced the parson with a disembodied eye.[36] In the twentieth century, in his interpretation of the diagram, C. K. Ogden dispensed with the human element entirely.[37] But even in these later diagrams—and the disembodied eye expresses this with great clarity—there is a need to understand relative position and point of view. The prepositions, in this respect, exemplify an important aspect of language in Wilkins that tends to disappear in accounts focused on his tables. Something else could stand in for the little man, but his function was important.

What to make of all of this from the standpoint of the history of diagrams and the history of maps? First, looking hard at Wilkins's diagram of prepositions clarifies just how challenging Wilkins's project was in its time. There were few precedents for the symbolic mapping that Wilkins was attempting. The sorts of grammatical diagrams that we reflexively associate with language pedagogy today were later developments, and some of these relied in important ways on Wilkins's precedent. Meanwhile, the visualizations of language that Wilkins did have available to refer to—ranging from allegories to trees to tables—were strikingly different from what he was attempting. Second, Wilkins's diagram reminds us that representations of space are inherently relational—this is one of the major points of Wilkins's diagram—and this fact is baked into language. As Lydia Liu has suggested, when Ogden evacuated the human figure from his version of Wilkins's diagram, he was explicitly attempting to portray language as a kind of immaterial *technology*, and more specifically, as a kind of

code.³⁸ This trajectory was already present in Wilkins, though, as is evident in the comparison of his diagram and Ogden's, he was not yet quite prepared to invite his reader into the machine in the manner of Ogden or of Foucault in the guise of Velázquez.³⁹

All this said, it is a strange world that Wilkins's parson inhabits, with direction but no perspective, and no ground for the ground. In the diagram, Wilkins's parson looks real enough. But the vista he looks upon does not. He faces across the page incongruously in the direction of two concentric circles shaded like planetary orbs in an astronomy text. These float in a non-space extending above and beneath him. And, of course, everything is covered with words.

It is *possible* to imagine a real-world scene looking something like the one in which Wilkins's man finds himself. He might be standing on a cliff edge looking at a very large floating orb.⁴⁰ But this is not what Wilkins is after. Nor is the visual relationship that Wilkins constructs between the human figure and the diagrammatic orbs allegorical or decorative in the manner of the portraits in Renaissance astronomy books, say. In those, the human figure provides an explanatory context for the diagram from a visual space distinct from the main diagram itself. Here, man is diagram.

Wilkins also clearly differentiates his diagram from cosmic diagrams of an allegorical sort published by contemporaries such as the physician Robert Fludd, who sought to convey a Paracelsian conception of man as microcosm of universal order. Wilkins makes this distinction explicit in the 1654 *Vindiciæ academiarum*, in which he and his collaborator, Seth Ward, first sketch some of the ideas realized fourteen years later in the *Essay Towards a Real Character*.⁴¹ In relation to the kind of human figure that we find in Fludd, we may say that Wilkins's diagram is *decentering*. The human figure in it is less the subject of the map than its compass rose. It is important to understand this point fully. Wilkins's little man makes neither an ontological nor a phenomenological claim. He is not the center of the universe. He is not our lens on the universe. He is a directional pointer. A key. A legend.

To give this claim a graphic point of reference, consider this compass rose from a sixteenth-century portolan atlas presented to Henry VIII in 1542, the *Boke of Idrography* by the French navigator Jean Rotz, which does in an elegant and artisanal way what Wilkins does in a telegraphic and philosophical one by employing a pointing human figure as a directional indicator in the center of its compass (fig. 6.7).⁴² Seeing Wilkins's diagram in relation to an artifact such as this also suggests the ways in which, though abstract and placeless, Wilkins's diagram relies on a common understanding of an embodied world and corporeal agency, in this case that of navigation, without attributing any specific philosophical priority to either.

Consider again the little parson standing with his left arm raised and his right posed on his hip, in a position that gesture manuals refer to as "general address," though Wilkins's speaker uses his left hand, where the gesture manuals generally counsel using the right. From this starting point, in combination,

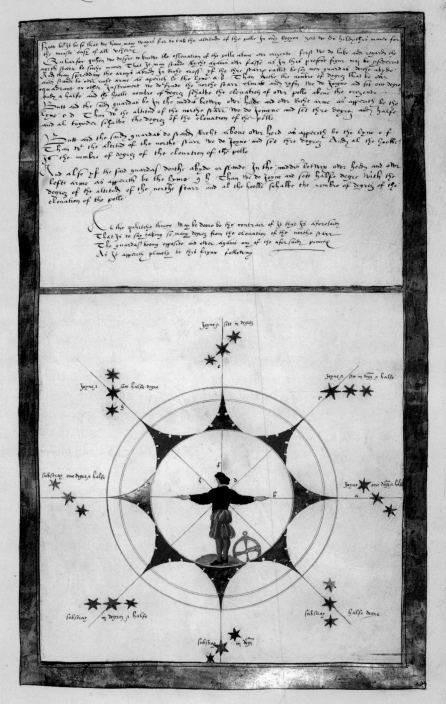

FIG. 6.7 Jean Rotz, compass rose from *Boke of Idrography*. Manuscript page (1542). British Library / Granger. All Rights Reserved.

small alterations of position and thoughtful timing with speech can create a near infinity of effects. Albert Bacon's 1881 *Manual of Gesture* lays out some of the possibilities offered simply by lowering the oblique supine hand, including as the first examples: "These are the FUNDAMENTAL principles of knowledge"; "These things are CERTAINLY true."[43] Not a bad start.

In general, Wilkins's diagram may remind us of others delineating rhetorically meaningful positions of the human body, whether in relation to itself or to a surrounding, abstract space as depicted in Gilbert Austin's *Chironomia, or, a Treatise on Rhetorical Delivery* of 1806, or in a more recent image of a person steering a Virtusphere virtual reality controller (fig. 6.8).[44] In the latter, the projection of space is made literal by the computer interface. Tilting one's head up brings the virtual things that are coded as above into view in the goggles. Stepping forward takes the floating I to and through and beyond whatever is projected before it. In virtual reality, we are in a telling space relative to the overall issues posed here. What we have there is not so much a map as a tool for exploration. In light of this, we can and should notice the insistence of the body-as-pointer in so many of these images. This is deictic, not Vitruvian, man.

We can also of course establish useful relationships between Wilkins's diagram of language and many kinds of scientific diagrams from his period. And Wilkins, at the center of the English scientific world, was well positioned to be aware of new visual conventions. In addition to his own researches on subjects as diverse as measurement standards, blood transfusion, perpetual motion, and harpoon and carriage design, he associated closely with most of the important English scientists of the day, including graphic pioneers such as Robert Hooke. Hooke's *Micrographia: or some Physiological Descriptions of Minute Bodies made by Magnifying Glasses* (1665), which included many beautiful plates depicting views through a microscope, pushed spatial thinking into new dimensions.[45] Given his close involvement in Hooke's project, it is clarifying to observe that Wilkins's diagram does *not* at all resemble the famous views through the microscope depicted there.

In fact, Wilkins's diagram of prepositions also differs from *his own* previous drawings, which offer no direct precedent for the kind of conceptual project we find in the diagram of local prepositions. There are, however, interesting details in Wilkins's earlier works on problems such as force and motion that employ deictic devices not unrelated to the parson, including floating hands and heads in his *Mathematical Magick: or, The Wonders That May Be Performed by Mechanical Geometry* (1648) (fig. 6.9).[46] The subjects of those images are *devices*, which are depicted simply, but because they are sometimes hypothetical and always functional, the diagrams include a kind of semaphore of movement in space, which breaks the unity of the picture and makes it speak.

So, here is at least a family resemblance: Wilkins's diagrams of devices and his diagram of language are both *engineering* diagrams. While there are analytic elements to the diagram of prepositions, we must remember that despite the convenient English words on it, it is not a diagram of English, it is a diagram of

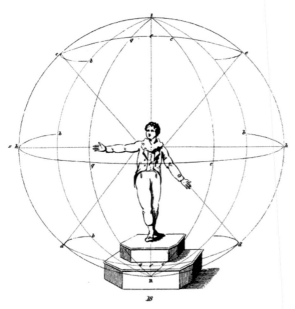

FIG. 6.8 Gilbert Austin, diagram of rhetorical position, from *Chironomia, or, a Treatise on Rhetorical Delivery* (1806); and Virtusphere virtual reality controller by Virtusphere, Inc., Binghamton, NY, with permission.

FIG. 6.9 John Wilkins, diagram of the wheel, by multiplication of which it is easie to move any imaginable weight, from *Mathematical Magick: or the Wonders That may be Performed by Mechanical Geometry* (1648). Library Company of Philadelphia.

his *invented* language, and it may be for this same reason that Wilkins, unlike contemporary grammarians, was able to imagine a schematic language diagram in the first place. Engineering is a kind of making, and the further we pursue the question, the clearer it becomes that Wilkins conceived his language as a kind of machine and his diagrams, like his book in general, as instruction manuals.

Another way of thinking about diagrams in Wilkins would be to borrow from John Bender and Michael Marrinan to observe the ways this kind of representation resists *imitation*: the diagram is powerful because it functions according to a logic intrinsic to its own organization. *Above* in the diagram expresses aboveness in relation to the other terms present and to the little man who orients us. The diagram in its totality demonstrates that the system only functions *as a system*. *Above* in Wilkins's diagram, for example, only has meaning in relation to the diagram's equally relational *below*. The same goes for *before* and *after*. Here, then, is a second role for the parson: in addition to orienting us to the system, he also figures us inhabiting it, if somewhat uneasily and not fully self-aware. Bender and Marrinan suggest that we think about a diagram as a "working object." They write:

> We take diagrams to be visual forms of description that make few concessions to imitation, meaning by "imitation" a staging of content as if belonging to a world both contiguous with and similar to our own. Our view of diagrams aligns them functionally with the "working objects" of late-eighteenth-century atlases discussed by Lorraine Daston and Peter Galison. "If working objects are not raw nature," they write, "they are not yet concepts, much less conjectures or theories; they are the materials from which concepts are formed and to which they are applied."[47]

Of course, here Bender and Marrinan are speaking specifically of the eighteenth-century *Encyclopédie* of Diderot and d'Alembert, in which the diagrams—other than the famous conceptual tree in the "Preliminary Discourse"—are mostly of things rather than concepts. Yet for some reason, the comparison is telling.

The visual diagrams of the *Encyclopédie*, argue Bender and Marrinan, structure the reader's experience of the visual plane. On the one hand, the plates evince a strong conceptual unity: out of heterogeneous elements, they present a functional assemblage. The most innovative plates of the *Encyclopédie* are not representations of nature—many of these are dependent on images from Wilkins's own period—but of the arts and industries. Here, as historian Ray Birn has shown, the *Encyclopédie* makes social and political as well as epistemological arguments.[48] More importantly for our purposes, the *Encyclopédie* also constructs arguments about itself as an *information system*, as suggested by Roland Barthes in his essay "Les planches de l'*Encyclopédie*" in 1964. Barthes writes:

> We can even specify more clearly what the man of the Encyclopedic image is reduced to—what is,

in some sense, the very essence of his humanity: his hand. In many plates (and not the least beautiful), hands, severed from any body, flutter around the work (for their lightness is extreme); these hands are doubtless the symbol of an artisanal world (again we are concerned with traditional, virtually unmechanized trades, the steam engine is kept out of sight), as is seen by the importance of the tables (huge, flat, well lighted, often encircled by hands); but beyond artisanship, the hands are inevitably the inductive sign of the human essence.... Hence in the immediate state of its representations, the *Encyclopedia* is constantly concerned to familiarize the world of objects (which is its primary substance) by adding to it the obsessive cipher of man. Yet beyond the letter of the image, this humanization implies an intellectual system of an extreme subtlety: the Encyclopedic image is human not only because man is represented in it but also because it constitutes a structure of information.[49]

The plates of the *Encyclopédie* achieve this specific information effect by flattening, dividing, and reconstituting space in such a way as to define an epistemological space at once self-sufficient and yet decomposable, socially situated and yet silent.

Diagrams in both the *Essay* and the *Encyclopédie* exhibit what Bender and Marrinan elsewhere refer to as a *hybrid* organization (fig. 6.10).[50] In the plates of the *Encyclopédie* and in related works from the period, the hybrid structure is vertical (scene at the top, elements at the bottom), whereas in Wilkins, it

FIG. 6.10 Diagram of needle making from Denis Diderot and Jean le Rond d'Alembert, *Encyclopédie, ou Dictionnaire raisonné des sciences, des arts et des métiers* (1751–1772). Library Company of Philadelphia.

is horizontal (scene at the left, elements at the right). The function of the hybridization is similar. By breaking up the pictorial unity of the space, the diagram establishes protocols for reading images as information. In Wilkins, we can go further still: it is not just a matter of coding the visual object as information; here, it is a matter of coding the space of the representation as itself the representation's object.

A Map of Language [141]

FIG. 6.11 Kellom Tomlinson, diagram of the regular order of the minuet, from *The Art of Dancing Explained by Reading and Figures* (1735). Library Company of Philadelphia.

The structural effect that Barthes identifies in the plates of the *Encyclopédie* is generated by the juxtaposition of two kinds of visual vocabulary, one scenographic and the other paratactic. At the top of the plate, we often find a tableau of the subject matter in use. Below, we find the elements of the tableau extracted into a kind of void. Wilkins's entire project is, of course, paratactic: it produces one giant list of words. Yet in his diagram, the abstract arrangement of space is all. In contrast to the plates of the *Encyclopédie*, the elements of this plate from Wilkins cannot be extracted and still retain their visual value, since relative spatial situation is precisely what the diagram depicts. In Wilkins, the visual interstices matter more than any things that may be represented: or, perhaps better put, they *are* the things represented.

The problem of how to code spatial relations is not limited to language; to the contrary, as Wilkins's diagram suggests, the graphic representation of spatial relations is in some ways more basic than the verbal. This is also an argument made explicitly by the Encyclopedists about their own plates. In those, and especially in representations of artisans at work, the *Encyclopédie* deftly portrays relationships and practices that can be expressed in language only partially and with effort.

A compelling case for this is the diagramming of choreography. In depictions of dance in the seventeenth and eighteenth centuries, human figures are common (fig. 6.11). Their function is ambiguous,

however. In some cases, they communicate body position; in some cases, scale. In the latter, they serve principally as a gauge of the choreography, not as the subject matter of the image. They depict only the instrument through which the choreography will eventually be actualized. In this sense, the human figures in the diagram are avatars of code: they indicate that the choreographic diagram is not a descriptive map but a set of instructions.[51]

The tension between the abstraction of such dance notation and the realism of the embedded human figures is strongly reminiscent of the tension at the heart of Wilkins's diagram of local prepositions. There, it arises from the strange way that the human figure inhabits a world of signs. All of this is part of the diagram's argument, telling us that what is depicted in it is space, not place, and—crucially, but also elusively—that the things depicted in the diagram are relationships. In Wilkins, the hybridization of figuration and graphic abstraction recodes the visual field. It invites multiple views; and it establishes a new protocol for the informational use of visual representation. At the same time, it offers a view of the symbolic tools preliminary to mapping as an exploratory act.

Notes

Research for this paper was supported by the William Andrews Clark Memorial Library at UCLA and by the Max Planck Institute for the History of Science, Berlin. Earlier versions were presented at the David Rumsey Map Center at Stanford University and the Renaissance and Early Modern Forum at the Hebrew University of Jerusalem.

1. Michel Foucault, *The Order of Things: An Archeology of the Human Sciences* (New York: Vintage, 1970), 74.

2. John Wilkins, *An Essay Towards a Real Character and a Philosophical Language* (London: 1668).

3. Wilkins, *Essay*, "Epistle to the Reader," ii.

4. Wilkins, *Essay*, 19: "The first thing to be provided for in the establishing of a Philosophical Character or Language, is a just enumeration of all such things and notions to which names are to be assigned." Depending on how you count, Wilkins's basic terms are between 2,030 (Umberto Eco) and 4,194 (Rhodri Lewis). Umberto Eco, *The Search for the Perfect Language* (New York: Blackwell, 1995), 245; and Lewis, "The Same Principle of Reason: John Wilkins and Language," in William Poole, ed., *John Wilkins (1614–72): New Essays* (Leiden: Brill, 2017), 192–94. For comparison, the second edition of the *Oxford English Dictionary* defines over one hundred seventy thousand words. "How Many Words Are There in the English Language?," Oxford Dictionaries, https://en.oxforddictionaries.com/explore/how-many-words-are-there-in-the-english-language/ (accessed Sept. 28, 2018).

5. Joseph-Marie Degérando, *Des signes et de l'art de penser: considérés dans leurs rapports mutuels* (Paris: Goujon fils, 1800), vol. 4, 399–400. See also Eco, *The Search for the Perfect Language*, 258; and Mary M. Slaughter, *Universal Languages and Scientific Taxonomy in the Seventeenth Century* (Cambridge: Cambridge University Press, 1982), ch. 4.

6. Wilkins's category system had a number of unforeseen applications, most famously its influence upon the thesaurus of Peter Mark Roget (1805). Barbara Shapiro, *John Wilkins 1614–1672: An Intellectual Biography* (Berkeley: University of California Press, 1969), 222.

7. John Wilkins, *Mercury: The Secret and Swift Messenger* (London: 1641).

8. Shapiro, *John Wilkins*, 213.

9. Jorge Luis Borges, "The Analytical Language of John Wilkins," in *Other Inquisitions, 1937–1952*, trans. Ruth L. C. Simms (New York: Simon & Schuster, 1968), 101–5. Borges's essay was first published in the Argentine newspaper *La Nación* on February 8, 1942.

10. Wilkins, *Essay*, 102, 127, 143.

11. For example, Wilda Anderson, *Between the Library and the Laboratory: The Language of Chemistry in Eighteenth-Century France* (Baltimore: Johns Hopkins University Press, 1984).

12. Wilkins, *Essay*, 415.

13. Wilkins, *Essay*, "Epistle": ". . . which being Philosophically unfolded, and rendered according to the genuine and national importance of Words, will appear to the inconsistencies and contradictions. And several of those pretended, mysterious, profound notions, expressed in great swelling words, whereby some men set up for reputation, being this way examined, will appear to be, either nonsense, or very flat and jejune."

14. Wilkins, *Essay*, 5.

15. Foucault, *Order of Things*, xv–xviii.

16. Borges, "Analytical Language," 104.

17. Wilkins, *Essay*, 270.

18. Wilkins, *Essay*, 263.

19. Wilkins, *Essay*, 263.

20. Wilkins, *Essay*, 453: "Now for the Latin Grammar, it doth in the common way of Teaching take up several of our first years, not without great toyl and vexation of the mind, under the hard tyranny of the School, before we arrive to a tolerable skill in it. And this is not chiefly occasioned for that great multitude of such Rules as are not necessary to the Philosophy of speech, to gather with the Anomalisms and exceptions that belong to them; the difficulty of which may well be computed equal to the pains of Learning one third part the words; according to which the labour required to the attaining of the Latin, may be estimated equal to the pains of Learning forty thousand words."

21. Wilkins, *Essay*, 20.

22. Eco, *Perfect Language*, 258–59.

23. Liu writes: "To Ogden, the architect of Basic, the practicalities of simplified language learning for international commerce and politics were only some of the reasons but not the only ones for promoting Basic. He saw simultaneous advances in a number of statistical domains as the immediate theoretical impetus of the project." From a practical point of view, Ogden's implementation of this kind of nesting is embodied in his so-called "panopticon," a language wheel demonstrating how words in Basic English may be meaningfully organized by a mechanical device. As Liu argues, Ogden's panopticon resembles nothing so much as the Lagado writing machine from *Gulliver's Travels*. "This is by no means an arbitrary comparison because prior to the invention of the computer, Ogden's statistical treatment of the vocabulary already presupposed a technological view of language." Lydia Liu, *The Freudian Robot: Digital Media and the Future of the Unconscious* (Chicago: University of Chicago Press, 2011), 90–1, 93.

24. See Michel de Certeau, *The Practice of Everyday Life* (Berkeley: University of California Press, 1984), 91–130.

25. Wilkins, *Essay*, 311.

26. John Horne Tooke, *Epea Pteroenta, or, the Diversions of Purley*, 2 vols. (London: J. Johnson, 1786, 1805), vol. 1, p. 249: "[T]he names of all abstract relation (as it is called) are taken either from the adjectives common names of objects, or from the participles of common verbs. . . . Wilkins seems to have felt something of this sort, when he made his ingenious attempt to explain the local prepositions by the help of a man's figure. . . . But confining his attention to ideas (in which he was followed by Mr. Locke) he overlooked the etymology of words; which are their signs and in which their secret lay."

27. C. K. Ogden, *Basic English: A General Introduction with Rules and Grammar* (London: K. Paul, 1932).

28. Horne Tooke, *Winged Words, or the Diversions of Purley*, 2nd ed. (London: J. Johnson, 1798), pt. 1, 454.

29. Wilkins, *Essay*, 309–11. On the seventeenth-century visual repertoire, see Veronica Della Dora, in this volume.

30. For a contemporary account of the analogy between space and time, see George Lakoff and Mark Johnson, *Metaphors We Live By*, rev. ed. (Chicago: University of Chicago Press, 2003).

31. Daniel Rosenberg and Anthony Grafton, *Cartographies of Time: A History of the Timeline* (New York: Princeton Architectural Press, 2010).

32. For example, Joseph W. Wright, *A Philosophical Grammar of the English Language* (New York: Spinning & Hodges, 1838), 76.

33. On time, see Brantley York, *An Analytical, Illustrative, and Constructive Grammar of the English Language* (Raleigh: W. L. Pomeroy, 1860), 82. For more early grammatical diagrams, see the website Coffee & Donatus: Early Grammars and Related Matters of Art and Design, http://www.coffeeanddonatus.org.

34. There is a long allegorical tradition of depicting the study of grammar as a component of the trivium; for example in the 1503 textbook *Margarita Philosophica* by the German humanist Gregor Reisch and the 1658 *Orbis sensualium pictus* by the Bohemian educational theorist John Amos Comenius. Grammar is treated abstractly in knowledge trees; for example, in the 1587 *Tableaux accomplis de tous les arts libéraux* by the French humanist Christophe de Savigny. An interesting variation on both themes is the allegorical figure of the grammar master depicted with his disciplining switch, which appears in mnemonic systems, including a remarkable educational card game called the *Logica memorativa* published by Thomas Murner in 1509. Around the time of Wilkins, one begins to see what Walter Ong calls "quasidiagrammatic" representations of the elements of grammar, as in the allegorical tableaus of Johannes Buno in his 1651 *Neue Lateinische Grammatica in Fabeln und Bildern*. Like Murner's game, Buno's project was more mnemonic than analytical. See also Walter J. Ong, "System, Space, and Intellect in Renaissance Symbolism," *CrossCurrents* 7, no. 2 (Spring 1957): 123–36; and "From Allegory to Diagram in the Renaissance: A Study in the Significance of the Allegorical Tableau," *Journal of Aesthetics and Art Criticism* 17, no. 4 (June 1959): 423–40. On the "tabularization" of grammar in early modernity, see Ray Schrire, "Shifting Paradigms: Materiality, Cognition and the Changing Visualization of Grammar in the Renaissance," *Renaissance Quarterly* (forthcoming).

35. Michel Foucault, *This Is Not a Pipe* (Berkeley: University of California Press, 1982).

36. William Casey, *Gramatica Inglesa para uso de los españoles*, 2nd ed. (Barcelona: Piferrer, 1827), folding chart. See http://www.coffeeanddonatus.org.

37. Ogden, *Basic English*.

38. As Eco explains, Leibniz explicitly articulated the possibility of a symbolic machine. And as Liu points out, Jonathan Swift did satirically as well. Eco, *Perfect Language*, 269–92; and Liu, *Freudian Robot*, 40, 93.

39. On the 1656 painting *Las Meninas*, see Foucault, *Order of Things*, 16: "Perhaps there exists in this painting by Velázquez, the representation as it were of Classical representation, and the definition of the space it opens up to us. And, indeed, representation undertakes to represent itself here in all its elements, with its images, the eyes to which it is offered, the faces it makes visible, the gestures that call it into being. But there, in the midst of this dispersion which it is simultaneously grouping together and spreading out before us, indicated compellingly from every side, is an essential void: the necessary disappearance of that which is its foundation—of the person it resembles and the person in whose eyes it is only a resemblance. This very subject—which is the same—has been elided. And representation, freed finally from the relation that was impeding it, can offer itself as representation in its pure form." As Lydia Liu has pointed out, the whole problem takes on new and intriguing dimensions in Ogden's period, some of which are developed in a paper by Ogden's partner in *Basic English*, I. A. Richards, entitled "Communication Between Men: The Meaning of Language," delivered at the 1951 Macy Conference on cybernetics, the same meeting at which Claude Shannon demonstrated his famous mechanical maze-solving mouse. Liu, *Freudian Robot*, 84–97.

40. There is a scene in one of the Asterix comics that looks just like this; there, the floating orb is a spaceship visiting Gallia Romana. Albert Uderzo, *Asterix: Le ciel lui tombe sur la tête* (Paris: Albert René, 2005).

41. John Wilkins and Seth Ward, *Vindiciæ academiarum containing some briefe animadversions upon Mr Websters book stiled, The examination of academies* (Oxford: 1654), 46. See also Shapiro, *John Wilkins*, 207.

42. Jean Rotz, *Boke of Idrography*, British Library, Royal MS 20 E. IX, fol. 4, the Rotz Atlas (1542). See also Michael Wintroub, *The Voyage of Thought: Navigating Knowledge across the Sixteenth-Century World* (Cambridge: Cambridge University Press, 2017).

43. Albert Bacon, *A Manual of Gesture; Embracing a Complete System of Notation Together with the Principles of Interpreta-*

tion and Selections for Practice (Chicago: S. C. Griggs, 1881).

44. Gilbert Austin, *Chironomia, or, a Treatise on Rhetorical Delivery* (London: T. Caddell and W. Davies, 1806). Virtusphere VR controller, by Virtusphere, Inc., Binghamton, NY.

45. In *Micrographia*, which appeared just three years before Wilkins's *Essay*, Hooke testified, "there is scarce any one Invention, which this Nation has produc'd in our Age, but it has some way or other been set forward by [Wilkins's] assistance." And in the case of his own book, this was precisely the case. Shapiro, *John Wilkins*, 201–3.

46. John Wilkins, *Mathematical Magick: or, The Wonders That May Be Performed by Mechanical Geometry* (London: 1648).

47. John Bender and Michael Marrinan, *The Culture of Diagram* (Stanford, CA: Stanford University Press, 2010), 33; and Lorraine Daston and Peter Galison, "The Image of Objectivity," *Representations* 40 (Autumn 1992): 85.

48. Ray Birn, "Words and Pictures: Diderot's Vision and Publishers' Perceptions of Popular and Learned Culture in the *Encyclopédie*," in Marc Bertrand, ed., *Popular Traditions and Learned Culture in France from the Sixteenth to the Twentieth Century* (Stanford, CA: Anma Libri, 1985).

49. Roland Barthes, "The Plates of the *Encyclopedia*," in *New Critical Essays* (Berkeley: University of California Press, 1980), 28–29.

50. Bender and Marrinan, *Culture of Diagram*, 42. Compare here the discussion of engineering diagrams in John Law, *Aircraft Stories: Decentering the Object in Technoscience* (Durham, NC: Duke University Press, 2002); and compare the discussion of analogies between dance and language in Susan Leigh Foster, *Reading Dancing: Bodies and Subjects in Contemporary American Dance* (Berkeley: University of California Press, 1986).

51. Compare in this volume Barbara Mundy's account of the world-making instructions—walk boundaries, shoot arrows, and so on—in Aztec and Mixtec maps.

PART III

The United States

Among the many tasks awaiting the American revolutionaries of 1776 was to craft a new cartographic regime for their young nation. From the cacophonous diversity of colonial and imperial maps that characterize the pre-1776 era now emerged a new genre: the map of the United States.

Today we take this genre in stride. Who has not used such a map? This section reminds us that the map of the United States is anything but an obvious project. Instead, it emerged at a particular moment in history. In a hostile international diplomatic context, as Susan Schulten's essay in this section shows, these maps projected onto the page a cheerful vision of the inevitability of the infant republic's expansion into a transcontinental future. The muscular project of American nationalism, she reveals, has always been highly cartographic because it has so often been imagined in spatial terms. Frederick Jackson Turner's famous "Frontier Thesis" of 1893 tied the story of the nation to its growth through time and space. The Turner thesis in turn reveals a larger truth: that the time element in US maps has often been futurity—spaces to be settled, conquered, absorbed, engrossed, or erased.

But still the past tugged on American maps, even in this future-oriented nation. Caroline Winterer's essay shows how nineteenth-century Americans gradually cast off the hated "New World" moniker that European elites had long deployed to argue that American life forms were shriveled novelties, inferior versions of their European betters. Over the course of a single century, Americans rebutted these arguments in part by using geology to show that the lands of the United States were in fact the oldest on the planet. They then used these facts to baptize their new national parks in the warm glow of an awe-inspiring primordial antiquity. James Akerman's chapter takes in a far shallower yet still critical past for the young nation by examining what he calls "time traveling": travel directed at developing insights about past events (in this case, major American wars) by visiting the terrain on which they unfolded. He shows how American battlefield tourist maps had to conform to a succession of new transportation technologies, each of which changed a person's experience of movement through time and space. Marking maps of the Gettysburg battlefield for foot travel was an entirely different problem than marking it for the speed

and isolation of an automobile.

All three essays suggest further lines of research so we can better understand how American maps have long quietly transcended the dominant westward-frontier-progress narrative. The essays in this section show that many American maps also recognized other, idiosyncratic temporalities that make up the human experience. From Winterer's "stream of consciousness" diagram, which tried to show the dissolving boundaries between a person's successive thoughts; to the surging population densities in Schulten's western maps; to the human-centered cyclorama of the Battle of Gettysburg from Akerman's essay: all show that other rich and compelling temporalities lurked behind the ideology of territorial progress glorified in Turner's 1893 address, "The Significance of the Frontier in American History." With so much digital history focusing on the United States, these essays are a call to action for future researchers to consider other temporalities.

7

Caroline Winterer

The First American Maps of Deep Time

Among the most transformative new ideas of the nineteenth century was deep time. Between roughly 1800 and 1900, Americans and Europeans began to argue that the earth's history extended millions and even billions of years into the past, far longer than the roughly 6,000-year age of the earth suggested by a literal reading of the Bible. Deep time was born in the young discipline of geology, but quickly migrated to biology, linguistics, anthropology, archaeology, paleontology, psychology, literature, and art. A mere hunch in 1800, deep time had utterly revolutionized thought a century later.[1]

Yet for all its popularity, deep time was also exceedingly difficult to conceptualize. How could the mind grasp the idea of thousands of years, let alone millions? Again and again, early proponents of the long chronology marveled at the human brain's inability to grasp such a lengthy span of time. One American geologist observed that the early earth was "a period so remote as to defy human conception," while another deemed the time span "incalculable."[2] British contemporaries also complained that the early earth was "inconceivably remote," a time "incomprehensibly vast" that "the imagination in vain endeavours to grasp."[3]

Here was a dilemma: a major new concept that could not be conceived. Deep time thus presented a new cognitive problem, different from the long-standing Christian idea of eternity. Christian theologians had long argued that God existed outside of space and time: the divine eternity was both unseen and unchanging. This idea was so foreign to human beings' lived experience of a visible and changing world that eternity became one of the attributes that made the divine deliciously mysterious and incomprehensible. By the seventeenth and eighteenth centuries, the new theory of a "state of nature" had quasi-naturalized eternity by locating a secular origin point for human societies. But the state of nature was widely understood to be a useful fiction. As Jean-Jacques Rousseau put it, the state of nature "no longer exists, perhaps never existed, and will probably never exist."[4]

By contrast, deep time was never a useful fiction. Proponents maintained that their hypothesis rested on the hardest facts of all: rocks. They pointed to cliff faces, fossils, and lava flows as proof that deep time was real. Rocks and fossils could be observed, measured, collected, and drawn. They not only documented the passage of eons, they seemed to reveal the

unfolding of universal laws of progressive historical development, from a more primitive then to a more advanced now, from a primordial world of creeping trilobites to a modern planet bustling with railroad-building human beings. Deep time rode on the new conviction that the world shared (or should share) a single, uniform, linear, progressive time.[5] Visible, material, and changeable, deep time was thus the opposite of the Christian eternity. Yet it was still so immense as to be as incomprehensible and awesome as the deity. One major task for the first proponents of deep time, then, was to find nontheological, naturalistic ways of explaining how the human mind comprehended the passage of quantities of time that far exceeded the range of lived human experience and perhaps even the capacities of the brain.

This essay examines three of those efforts in the fields of geology, psychology, and American landscape painting in the first flush of deep time's popularity. They do not strike us as an obvious group today. But they were all new disciplines in their day, and therefore among the most avant-garde modes of inquiry, confronting ideas about humanity, the earth, and God that were new and unsettling. They shared attitudes, sensibilities, metaphors, and visual strategies at a moment of great fluidity between what was map and non-map, art and non-art, science and non-science.[6] Their common quest for understanding led them to pioneer new kinds of images that mixed map, art, science, and text. In fact, they were some of the first people to use the new verb "visualize" to describe the effort to clarify abstract concepts with pictures.[7]

They also dared to face some of the most troubling implications of the lengthening calendar. The geological long chronology now displaced the Bible as the chief story of earthly change and rendered God nonessential to historical meaning. Deep time also seemed both empirically verifiable but also utterly beyond the human mind's ability to fully grasp it. Geologists, scientists, and landscape painters all wondered how the human brain and mind—now thought to have evolved over millions of years—apprehended time and operated in time. They confronted the problem of a brain that had evolved over deep time, but that for some reason was unable to fully grasp deep time. They lobbied for visualizations that would make plain the operations of time to minds whose time-grasping qualities they struggled to understand. The images of deep time that emerged in the nineteenth century had these concerns whirring in the background.

The first advocates of deep time were not entirely successful in making their new concept conceivable. They pioneered a whole gallery of diagrams, maps, paintings, and charts whose descendants we still use today. Yet, like those nineteenth-century scientists and artists, we continue to struggle with our inability to fully grasp deep time, even while accepting that our brains evolved over the course of it. Visualizations continue to proliferate, even as the near-infinity of deep time stubbornly resists total comprehension. This essay examines the first Americans to start down the difficult path we continue to walk today.

Deep Time in Earth Strata

We can begin among the geologists, since ideas about deep time emerged with them first. David Dale Owen, a geologist with a poet's heart, was among the first American thinkers to propose how to make deep time comprehensible to the human brain.

Born in 1807 and educated in Scotland, England, and Switzerland before sailing to the United States, Owen lived for most of his adult life in the town of New Harmony, Indiana. It had been purchased in 1825 by his father, the Welsh social reformer Robert Owen, who dreamed of creating a utopian community in which education and scientific inquiry could flourish for all. Although the Owenite social experiment quickly failed, the town of New Harmony remained a hub for progressive education and scientific research for decades after. Some of the most eminent scientists of the day stopped in, including the British geologist Charles Lyell and the entomologist Thomas Say. The intellectual hothouse of New Harmony remained David Dale Owen's home base during his long career leading some of the first geological surveys of the Midwestern states, launched by state governments hoping to capitalize on valuable coal, mineral, and ore deposits. From the 1830s until his death in 1860, Owen led the first geological surveys of Indiana, Kentucky, Arkansas, Wisconsin, Iowa, and Minnesota.[8]

Owen's surveys built on the emerging consensus that North America was a layer cake of time, a continent made of rock layers called *strata*. The terminology was imported from England and Scotland, where mineralogists and fossil hunters such as William "Strata" Smith promoted methodologies for the emerging science of geology that quickly caught on in North America.[9] As distinct from the two-dimensional timeline, in which events were represented as points, the stratum implied a three-dimensional bucket of time. Each rocky layer signified not a single event, but a process—what British geologist Charles Lyell called a vast "quantity of time" in which deposition had occurred.[10] In contrast to the morally meaningful human-centered events of recorded history, the strata therefore told an existentially disturbing story of an inconceivably long planetary past full of processes but largely empty of meaningful events beyond the great primordial flood (or, increasingly, floods) to which many geologists in Europe and America still subscribed.

David Dale Owen was personally influenced by the Scottish-born geologist William Maclure, one of the first people to imagine an America built of strata. The first professional geologist in the United States, Maclure was among the scientific luminaries who lived for a time in New Harmony. In 1809, he had published the first self-styled "geological map" of the United States, whose territory extended to the region slightly west of the Appalachians (fig. 7.1).

What made Maclure's map "geological" were the colored bands representing rocks of various ages, from the oldest ("primitive") to the newest ("alluvial"). Maclure insisted that the purpose of geology was to investigate the "relative position" of the rocks within the big picture so as to see "the great and prominent outlines of nature." To focus on isolated rocks would be like a "portrait painter dwelling on

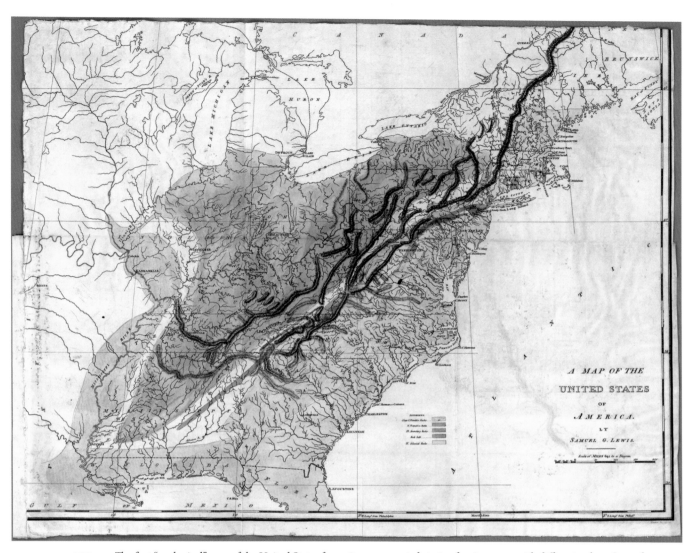

FIG. 7.1 The first "geological" map of the United States from 1809 represented strata of various ages with different colors. Samuel G. Lewis, "A Map of the United States of America." In William Maclure, "Observations on the Geology of the United States, explanatory of a Geological Map," *Transactions of the American Philosophical Society* 6 (1809): 411–28. Courtesy of the David Rumsey Map Collection, David Rumsey Map Center, Stanford Libraries, https://purl.stanford.edu/dk539gf3381.

the accidental pimple of a fine face." Maclure declined to speculate on "the relative periods of time" in which the rocks were formed and modified. "Such speculations are beyond my range," he declared.[11]

Maclure's map opened the door for an array of new maps that used rock strata as the basic structure of the map. The representation of time became essential rather than incidental to the map.[12] In Edwin James and Stephen Long's map of 1822 (fig. 7.2), which was inspired by Maclure, a knife has cut through the layer cake of North America.[13] The horizontal axis measures space (degrees of longitude west from Washington, DC). The vertical axis measures time (in feet from sea level upward, from the oldest to the newest strata). From the 1820s onward, the visual innovation of the so-called "section" quickly caught

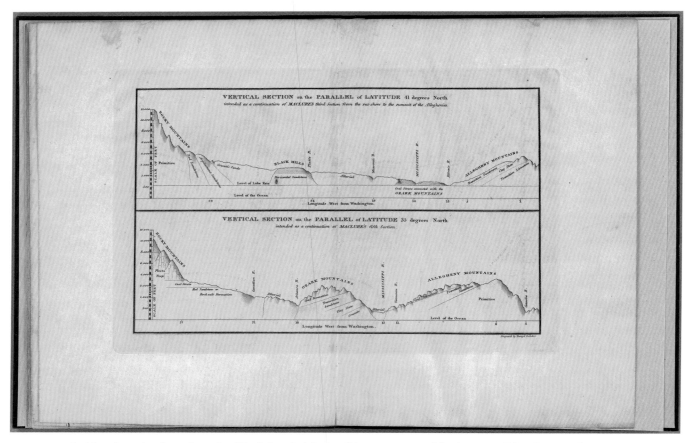

FIG. 7.2 In this early sectional map from 1822, North America's layers of time are uncovered for the viewer. Edwin James and Stephen Long, "Vertical Section on the Parallel of Latitude 41 Degrees North." In *Account of An Expedition from Pittsburgh to the Rocky Mountains, Performed in the Years 1819 and 1820 . . . Under the Command of Major Stephen H. Long* (Philadelphia: H. C. Carey and I. Lea, 1822). Courtesy of the David Rumsey Map Collection, David Rumsey Map Center, Stanford Libraries, https://purl.stanford.edu/by969kx4302.

on to represent deepening time in North America.

Prone to purple prose and awestruck digressions on the lifestyle of the rattlesnake, Owen wrote his geological reports with the idea that the human mind would be enriched by the contemplation of the "dark abyss beyond the human epoch." He hewed to a quasi-Transcendentalist belief that happiness lay in the contemplation of nature. Among all the natural sciences, he singled out geology as the most "sublime" because it made people aware of "remote periods" and "by-gone existences."[14] Fossils of "bygone ages" permitted humans "to contemplate, and restore to our perceptions, the very fishes, mollusks, and corals, that swarmed in the carboniferous seas millions of ages ago."[15] Such lengths of time were unknown a mere century ago, so Owen searched for new explanatory methods to make plain to the human mind "a period so remote as to defy human conception."[16]

Owen was not the only mid-century American pioneering new visualizations to make the long chronology more intelligible. The educator Emma Willard promoted her "Temple of Time" (fig. 7.3) as the best vehicle for clarifying the complexities of representing time in space.

The First American Maps of Deep Time   [153]

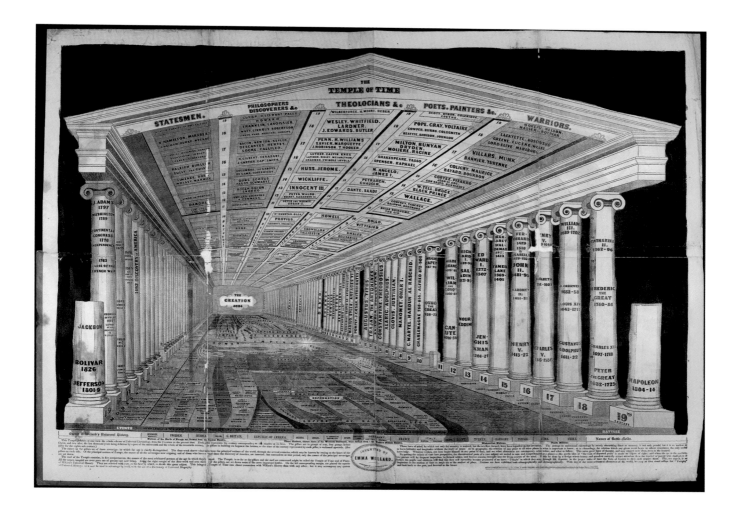

Willard described her Temple of Time. "The Chronographer, called the Temple of Time, or Universal Chronographer, is an invention by which time is measured by space, and all time since the creation of the world is indicated at once to the eye. By the unity of its plan, it can be easily understood, and remembered." She hoped that by "putting the Stream of Time into perspective, and adding light and shade," she could sew the idea of historical time "into the living texture of the mind."[17]

Willard herself rejected the new long chronology. She believed that the Old Testament was "the only valid authority concerning the creation of the world." Her Temple of Time gave literal form to that idea of a firm, known point of creation. On the far wall loom the words "The Creation 4004." All history flowed from this known beginning. Willard acknowledged that others might assign a "more distant date" to the Creation than 4,000 years before Christ. Those people could adapt the Temple of Time by extending it "further into the distance."[18]

And extend they did. While the moment of Creation was visible, knowable, and concrete in Emma Willard's temple, geologists such as Owen now coped with a moment of creation that had receded into uncertainty. Their challenge was to create new techniques to visualize the long calendar and its mysterious—and therefore invisible—beginning.

FIG. 7.3 In her "Temple of Time," the educator Emma Willard showed the history of the world since the Creation in a single image. Emma Willard, *Willard's Map of Time: A Companion to the Historic Guide* (New York: A. S. Barnes and Co., 1846). Rare Book Division, Department of Rare Books and Special Collections, Princeton University Library.

Owen explained the outline of a new visual method in 1843 before the Association of American Geologists and Naturalists in a paper entitled "Geological Paintings and Illustrations." Gone were Willard's architectural metaphors. Owen's method would be based on observation and measurement of the landscape itself, by geologists exposed for months on end to the harsh elements. Speed was therefore essential in the field. Owen explained that in a mere four months of geological fieldwork, he had painted "nearly eight hundred figures of organic remains, inclusive of lettering and stratification, and two large geological charts besides."[19] (By "painted," Owen meant not just painted by him, but also by a cadre of Native American informant-artists, who helped him find and correctly depict features of the landscape.)[20] He touted the many advantages of his method of water coloring: the paintings were cheap, quick, easily corrected, equally clear by candlelight and daylight, and easily transported.

In addition to these practical advantages, Owen's paintings compared favorably (in his opinion) to the "beautiful landscapes" of William Thompson Russell Smith, a Scottish-born American painter of mid-Atlantic landscapes.[21] Owen's descriptions of his surveying expeditions teem with comments about the picturesque beauty of the terrain. He frequently used the term "landscape" to describe what he saw, as though apprehending the surveys as an aesthetic and scientific project simultaneously. This recourse to imagination had been authorized by none other than Charles Lyell himself, in the first volume of his landmark *Principles of Geology* (1830). The human mind and spirit would be the great vehicles of discovery for vast times and spaces, argued Lyell. "Although we are mere sojourners on the surface of the planet, chained to a mere point in space, enduring but for a moment of time, the human mind is not only enabled to number worlds beyond the unassisted ken of mortal eye, but to trace the events of indefinite ages before the creation of our race." The human mind was "free, like the spirit."[22]

Owen's small "geological paintings and illustrations" appear in great number throughout his *Report of a Geological Survey of Wisconsin, Iowa, and Minnesota* (1852), the most ambitious geological survey undertaken to date in the United States (fig. 7.4). The report's purpose was primarily economic: to locate coal and mineral deposits for government exploitation. Conscious of his patrons, Owen apologized at the outset of the report for including any information of a "purely speculative" nature, as opposed to the "strictly practical and business" purpose of the survey.[23] But the long report frequently drifted away into reveries on the long chronology and the wonders of fossils. "The study of organic remains in rocks is, indeed, a most beautiful, a most fascinating research," it reads.[24] The little paintings showed how early images of deep time combined aesthetics and science.

Some of Owen's images broke new visual ground by marrying the new "sectional" view of ancient rocks to the existing genre of landscape painting. Owen's visualization of the Des Moines River in Iowa (fig. 7.5) exemplifies his attempt to capture the reality of the deepening chronology by experimenting with visual techniques.

FIG. 7.4 This is one of the many small "geological paintings" that dot Owen's 1852 geological survey. Many were created with information from Native Americans, some of whom are named in the text. "Castellated Appearance of Lower Magnesian Limestone, Upper Iowa." In David Dale Owen et al., *Report of a Geological Survey of Wisconsin, Iowa, and Minnesota: And Incidentally of a Portion of Nebraska Territory* (Philadelphia: Lippincott, Grambo, and Co., 1852), 65. Courtesy of the Department of Special Collections, Stanford University Libraries.

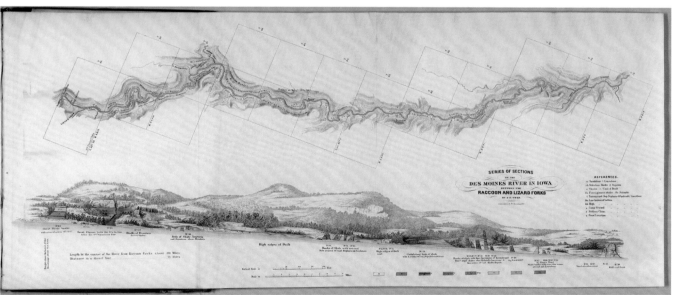

FIG. 7.5 This 1852 illustration of the Des Moines River represents time and space in multiple ways. "Series of Sections on the Des Moines River in Iowa, between the Raccoon and Lizard Forks." In David Dale Owen et al., *Report of a Geological Survey of Wisconsin, Iowa, and Minnesota: And Incidentally of a Portion of Nebraska Territory* (Philadelphia: Lippincott, Grambo & Co., 1852). Courtesy of the David Rumsey Map Collection, David Rumsey Map Center, Stanford Libraries, https://purl.stanford.edu/pj744tc9737.

As the title "Series of Sections of the Des Moines River" implies, this image combines at least four visual conventions: the "sectional" view (the riverbanks color-coded with geological strata); landscape painting (the hillside above the banks of the river); a bird's-eye view of the river at top bedecked with flowery description reminiscent of earlier mapping traditions ("beautiful prairie"); and finally, looming overhead, a traditional map marking the latitude and topography of the Des Moines River.

Not just space but time is represented from multiple angles. The viewer stands in the river's present, with its gently flowing blue water and its luxuriant trees. The colored strata on the steep banks of the river reveal a distant past "beyond the grasp of our conceptions."[25] This image is a hybrid of scientific and philosophical impulses, pulled in one direction by the practical economic demands to show the location of the coal seams, and in the other by the reveries of deep time.

Owen's conclusion, ventured cautiously in his own geological reports but with increasing bombast by other scientists over the next decades, was to suggest that America was older than Europe. Some strata in the American West were "in all probability, the oldest fossil-bearing rocks yet brought to light in any part of this continent, if not of the world."[26] Combining art and science, landscape and stratigraphy, wonder and practicality, and present and past, Owen's sectional view of the Des Moines River represents an early effort to hint at deep time. In the next section, we will see how Owen's contemporaries, a new breed of professional psychologists, approached the problem of comprehending the incomprehensible depths of the remote past.

Deep Time in the Human Brain

For psychologists, the long chronology introduced the new problem of explaining how the human mind had evolved over long expanses of time. From an eighteenth-century world in which the "primitive" mind of "savages" was separated from the civilized mind through a vague sense of progress through stages in an unspecified amount of time, students of the brain now had to cope with the geologists' hard reality that all life on earth may have changed over thousands and perhaps millions of years.

Eager to establish psychology as an empirical science based on facts rather than armchair speculation, psychologists and other theorists now asserted that the human brain and its crowning achievement—the self-conscious mind—had evolved over eons from more primitive forms of life. These lower forms persisted in the "lower" structures of the brain nestled low in the head, even as humanity's more abstract reasoning powers blossomed in the novelty of the "higher" regions riding aloft in the vaulting human skull. This terminology was simultaneously spatial and normative, implying progressive development upward from primitive antiquity to modern glory. But how did the mind apprehend the passage of time—especially deep time? For this project, thinkers devised new diagrams that hinged on the same presupposition of "strata" that earth scientists' diagrams did: that is, of expressing time in terms of space.

Psychologists' new preoccupation with time was spurred in part by geologists. In his *Principles of Geology* (1830), Lyell had promoted the acceptance of deep time as "the characteristic feature of the progress of the science." He sketched the scenario by which humanity's distant memories of catastrophic earth-shaping events might be transmitted to their superior modern heirs. Using the example of the "barbarous and uncultivated" tribes of South America described by the Prussian explorer and naturalist Alexander von Humboldt, Lyell showed how ancient ideas could still shape modern minds. "The superstitions of a savage tribe are transmitted through all the progressive stages of society, till they exert a powerful influence on the mind of the philosopher. He may find, in the monuments of former changes on the earth's surface, an apparent confirmation of tenets handed down through successive generations, from the rude hunter, whose terrified imagination drew a false picture of those awful visitations of floods and earthquakes, whereby the whole earth as known to him was simultaneously devastated." In a section called "Modern Progress of Geology," Lyell explained that human social progress enabled modern scientists to grasp deep time. "By the consideration of these topics, the mind was slowly and insensibly withdrawn from imaginary pictures of catastrophes and chaotic confusion, such as haunted the imagination of the early cosmogonists. Numerous proofs were discovered of the tranquil deposition of sedimentary matter and the slow development of organic life."[27]

The first major theorist to approach the problem of integrating the human brain into an evolutionary framework and explaining its comprehension of the passage of time was the British philosopher Herbert Spencer. A big-picture systematizer in the vein of Karl Marx and John Stuart Mill, Spencer strove to harness everything—the cosmos, the earth, all life, and human society—into one overarching idea: evolution. Well before Charles Darwin had published *The Origin of Species* (1859), Spencer was widely known as an evolutionist whose publications argued not just for the reality of evolution over deep time, but for evolution as progress, ascending ever upward toward a perfection embodied by modern European men. An effort to make psychology a science based on empirical evidence rather than "a mere aggregation of opinions," Spencer's *Principles of Psychology* (1855) applied evolution to the study of the human mind.[28] Spencer described the mind as a physical object that evolved over time from a primitive nervous bundle to the glorious complexity of the modern human brain. He distinguished between the "developed" mind of modern Europeans and the "undeveloped" mind of savages and ancient Hebrews, who lumbered along without the aid of progressive concepts.[29]

Because time and space were major concerns in Spencer's evolutionary model of the brain, he had to wrestle with Immanuel Kant's claim that the sense of time and space constituted a priori knowledge—that is, knowledge originating in the mind independently of sensory experience. In his *Critique of Pure Reason* (1781), Kant had argued that space and time did not exist as external realities. It was the mind itself that

imposed spatial and temporal understandings on a person's experience of the world. The mind was unable to conceive of there being no space, though it could grasp a space with nothing in it; nor could the mind imagine there being no time at all, though it could imagine empty time.[30]

Kant had lived in the eighteenth century, before the rise of deep time, so had obviously not addressed the question of how the mind's perception of time and space had evolved over time. But Kant's notion of an a priori conception of space and time troubled Spencer, and his *Principles of Psychology* attempted to refute it.[31] Against Kant's notion of a subjective space and time, Spencer insisted instead that time and space were real, and that the brain experienced that reality in a way that had evolved over time to become more sophisticated. Primitive people could imagine space only in terms of time (for example, the distance between here and there is three days' walk). By contrast, evolved moderns could imagine time in terms of space (the distance between the 5 and 6 on a clockface represents one hour). "[T]hese facts will be seen to possess considerable significance," wrote Spencer. "*That in early ages, and in uncivilised countries, men should have expressed space in terms of time, and that afterwards, as a result of progress, they should have come to express time in terms of space* [emphasis in original]."[32]

The idea of an evolving human brain led psychologists to the same metaphors of strata and layering to which geologists had also turned. By the late nineteenth century, the eminent British neurologist J. Hughlings Jackson was arguing that the nervous system formed an "evolutionary hierarchy."[33] Following the laws of organic evolution from homogeneity to heterogeneity, the brain was made of nervous strata that ascended from the lowest and least complex (such as the spinal cord) to the highest, most unstable, and most complex, the regions where consciousness reigned. A materialist, Jackson called these highest brain centers "the anatomical substrata of consciousness."[34] Jackson imagined that insanity was caused by the literal de-evolution of the brain from the unstable highest center to the nether regions, a term he called "dissolution."

Jackson's American contemporary Chauncey Wright pushed further still on the evolutionary front. Influenced by Darwin, Wright attempted to explain how that most human of capacities—self-consciousness, the self's awareness of the thinking self—could have emerged from the process of the organic evolution of the brain without having been predetermined by it. "[N]o act of self-consciousness, however elementary," he wrote, "may have been realized before man's first self-conscious act in the animal world; yet the act may have been involved potentially in pre-existing powers or causes."[35] Self-consciousness was a radically new use of older structures and powers, arising suddenly, dramatically, and discontinuously from the deep eons of the evolutionary past.

The American psychologist William James became similarly preoccupied with explaining the thinking mind in an evolutionary perspective. Like Jackson, whose nervous-stratum model of the brain he cited approvingly in his *Principles of Psychology*

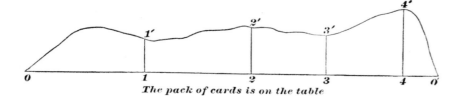

FIG. 7.6 William James's visualization of "the stream of consciousness" represented a thinking brain using similar visual techniques to the geological section. William James, *The Principles of Psychology*, 2 vols. (New York: Henry Holt, 1890), vol. 2, p. 279. Courtesy of the Department of Special Collections, Stanford University Libraries.

(1890), James lobbied for a "strictly positivistic" rather than metaphysical view of the brain. And like Jackson, James spoke of the brain as the "mechanical substratum" of thought, and called for a "complete diagram" of the brain.[36] He was influenced by George Romanes's *Mental Evolution in Man* (1888), which asserted that the human mind had evolved during "the abyss of planetary time."[37]

The brain might have evolved over deep time into a layer cake of nerves, but how could James explain the succession of individual thoughts that unfolded over time to constitute "thinking"? James called thinking "the first fact" of psychology and challenged himself to explain it empirically. To do so, he had to confront our sense that individual thoughts constitute an unbroken continuity. A person does not experience thinking as a series of separate brain states arranged like cars on a train, but rather as an unbroken flow of ideas. This caused James to liken thought to a stream, and he coined the phrase "stream of consciousness" or "stream of thought."[38] Though made up of innumerable temporary brain states, the stream of consciousness dissolved the breaks between thoughts, giving people the impression that their thoughts flowed continuously. The brain, like the earth, was built of hierarchical strata deposited over time. And like the earth, the mind had strata cut through by rivers over time.

To illustrate the stream of consciousness, James created a diagram that shows the brain thinking the sentence "The pack of cards is on the table" (fig. 7.6). The horizontal axis represents time, with each point marking an instant. The vertical axis represents the content of thought. Each of the resulting five strata in the diagram represents what James called "the object in the mind at the instant." James recognized that each instant, though represented as separate and distinct in the diagram, was in reality inseparable from all the others. As he put it, "they melt into each other like dissolving views." The diagram grew taller on the right, at the end of the sentence, because the final way of feeling the content of the sentence was "fuller and richer" than it was at the launching of the sentence.[39] That is, by the time you had thought the whole sentence, you were no longer thinking just of a pack, or even just of a pack of cards, but you were thinking that the pack of cards was lying on the table.

James had created a metaphor for cognition strikingly similar to the geological stratum. Both used the "section" (or "cross-section") of time and space to show processes unfolding in time. In the modern, evolved human brain, consciousness flowed like a stream, each individual thought unique and distinct yet ultimately connected to all the others to form a full thought. A cross-section of thinking revealed separate thoughts united in their general idea, in the way that a cross-section of rocks would reveal a "system" of characteristic fossils and rocks, as the British geologist Roderick Murchison had argued in his influential work, *The Silurian System* (1839).[40] "Annihilate a mind at any instant," explained James of his diagram, "cut its thought through whilst yet uncompleted, and examine the object present to the cross-section thus suddenly made; you will find, not the bald word in process of utterance, but that word suffused with the whole idea."[41]

Like geologists, psychologists attempted to understand the new and awesome temporal possibilities of deep time in the physical structures of rocks and brains. Both groups yearned for a new science based on what they called material "facts" and "data" rather than on metaphysical speculation. The evolution of humanity over deep time seemed to both groups to offer a plausible explanation for why only now, in the modern era, deep time was coming to be understood. Only the modern brain, evolving from simple savagery to complex civilization, could grasp how it understood the passage of time. Charts, diagrams, sections, and other images explained the evolved, modern, time-grasping mind to itself. The ability to self-consciously apprehend all sorts of time, including deep time, thus became the self-fulfilling explanation of the veracity both of deep time and of the evolutionary process that yielded the modern, civilized brain.

Deep Time and Landscape Painting

Deep time's visual appeal also extended to landscape painting. During the nineteenth century, landscape painting emerged as the archetypal American style of painting. Artists infused their depictions of nature with messages about the God-given historical importance of the United States and its transcendent religious meaning. Geology formed an important part of this nationalist program. American painters created giant canvases of awe-inspiring geological marvels such as Yosemite and the Grand Canyon. By the 1860s, scientists such as Harvard's Louis Agassiz were announcing that the "new world" was the oldest world of all, the "first-born among the Continents."[42] American landscape painters responded to the exciting imaginary of geology by depicting rock formations with scrupulous accuracy. One art critic at the time called these paintings "rock-portraits."[43]

But artists' new preoccupation with geology highlighted one of the eternally frustrating limitations of painting. While an image could capture discrete moments (as in a medieval triptych) or the sequential passage of time (as in William James's diagram of the mind thinking a sentence), it could not capture the fluid nature of time. In his *Laocoön* (1776), the German philosopher and art critic Gotthold Lessing had distinguished between the temporal and spatial arts. "Succession in time is the province of the poet, coexistence in space that of the artist."[44] By the second half of the nineteenth century, the telegraph and train had normalized speed and even long-distance simultaneity. Painters now felt even more acutely their inability to depict the flow of time, let alone movement itself.[45] The rock-portraits were born in this moment.

Martin Johnson Heade's 1862 painting *Lake George* (fig. 7.7) exemplifies the effect of deep time's discovery on the American landscape tradition. Heade was nationally known as a specialist in seascapes, shorelines, and marshes in the style of the Hudson River School.[46] He was one of many artists to paint Lake George, a popular New York tourist destination tucked into a fold in the Adirondack Mountains. Other representations included those by John W. Casilear (1857), Asher B. Durand (ca. 1860), John F. Kensett (several paintings from the 1860s

FIG. 7.7 This "rock-portrait" of Lake George in New York suggests a landscape shaped by geological forces. Martin Johnson Heade, *Lake George*. Oil on canvas. 1862. Courtesy of the Museum of Fine Arts, Boston, Bequest of Maxim Karolik 64.430.

and 1870s), and some lithographs by Currier and Ives. These artists depicted Lake George as a place of serene natural beauty and the site of battles in the Seven Years' War and Revolutionary War.

By contrast, Heade's *Lake George* was a postcard from deep time. Twice as wide as it is high, the panoramic view of the lake appears at first glance to be a typical rendering in the landscape mode. In a gigantic sky, late afternoon clouds melt over the low hills. The three boats hardly ruffle the steely lake. A closer look reveals that this is a rock-portrait. In fact, it is a family rock-portrait, each rock a unique individual that forms part of a larger group. A massive cracked slab of reddish rock juts in from the right, smooth as a tabletop. Striated boulders of various colors and sizes crowd its sides and top. Lichens cling to the square face of the large boulder in the right foreground, which rests with a smaller companion in this improbable location. The beach stretching to the left is a rock-strewn wedge interrupted by a few scrubby plants. Around a driftwood log, jagged stones poke up through the sand like tombstones.

The rock family portrait told a story of Lake George's ancient past. Heade painted the lake just as its geological history was coming into focus as the product of an ice age. In the 1840s, Louis Agassiz had proposed the controversial idea of a "glacial epoch" to explain the curious alpine features of his native Switzerland.[47] By the 1860s, the glacial theory was gaining international acceptance. There was now little doubt that all of North America "has been scarped, scraped, furrowed and scoured by the action of ice," as one proponent put it.[48] According to glacial theorists, massive sheets of ice had once blanketed much of the planet. As the glaciers smothering New York receded, they scoured the mountains to smooth nubs, dropped low piles of sand and boulders (moraines), trapped glacial meltwater behind those moraines to form pools such as Lake George, and randomly scattered large boulders in unlikely locations (glacial erratics). This glacial theory was controversial in part because it disrupted the reigning consensus that the earth was cooling over time from its birth as a fiery ball. The glacial theory opened the door to the unsettling idea that climate could fluctuate, cooling and warming and cooling again over eons. The modern conception of climate change (perhaps better termed climate variability) was born at this moment.

Louis Agassiz visited Lake George in the early 1860s, and like Heade was moved by the geological possibilities of its shoreline. After his visit, he published a popular essay entitled "The Silurian Beach," in which he imagined that the shores of Lake George showed the fossil footprints of a vanished world.[49] The term "Silurian," coined in the 1830s by the British geologist Roderick Murchison, designated one of the most ancient geological periods.[50] Shallow seas had then blanketed the earth, teeming with primordial marine creatures such as the crab-like trilobite, with its strangely modern eyes that looked out on the infant planet. What tiny strips of land poked up from the Silurian seas remained lifeless. The Silurian became a favorite subject of the first prehistoric landscape illustrators. One image from 1866 shows the divine rays lighting a Silurian beach littered with dead trilobites (fig. 7.8).

FIG. 7.8 Trilobites have washed ashore on this primordial beach of the Silurian Period. Louis Figuier, "Ideal Landscape of the Silurian Period," in *The World Before the Deluge. Containing Twenty-Five Ideal Landscapes of the Ancient World*. Translated from the French ed. (New York: Appleton, 1866), 92. Courtesy of the Department of Special Collections, Stanford University Libraries.

Agassiz was one of many American scientists to become excited about the Silurian period because they were finding armies of trilobites entombed in Silurian rocks in the United States. Here was further evidence that the New World might be older than the Old World. To Agassiz, the Silurian beach also offered a religious lesson. An opponent of Darwinian evolutionary theory, Agassiz instead supported the idea of multiple, separate creations. He urged his readers to appreciate the Silurian beach for the evidence it showed of God's tireless, sequential creativity. The whole of the early earth's dry land was merely a beach during the Silurian period, a fragile stage for the first land-dwelling plants and animals to emerge from the worldwide sea. Like Eden, the Silurian beach displayed God's endless benevolent energy in creating and recreating life on earth. "Let us remember, then, that, in the Silurian period, the world, so far as it was raised above the ocean, was a beach," counseled Agassiz, "and let us seek there for such creatures as God has made to live on sea-shores, and not belittle the Creative work, or say that He first scattered the seeds of life in meagre or stinted measure, because we do not find air-breathing animals when there was not fitting atmosphere to feed their lungs, insects with no terrestrial plants to live upon, reptiles without marshes, birds without trees, cattle without grass, all things, in short, without the essential conditions of their existence." God had lavished as much care on the Silurian trilobites as he had on modern humans. For what other reason would he have given them eyes, if not to behold the beauty

FIG. 7.9 This recent visualization of deep time shows the history of planet earth as an uncurling ammonite; a surfer marks the terminus of geological history to date. Joseph Graham, William Newman, and John Stacy, "The Geologic Time Spiral—A Path to the Past" (ver. 1.1), US Geological Survey General Information Product 58, poster, 1 sheet, 2008, http://pubs.usgs.gov/gip/2008/58/.

of their world? To fully appreciate the benevolence of the deity, modern people should carefully study the strange creatures of those early worlds, listening to their messages. "The crust of our earth is a great cemetery," wrote Agassiz, "where the rocks are tombstones on which the buried dead have written their own epitaphs."[51]

The enormous, detailed beach in Heade's *Lake George* offered a window onto deep time. The tombstone rocks reminded viewers of a vanished world, perhaps a world of ice, or perhaps an earlier world still, of trilobites washed onto sunlit beaches lost in the abyss of antiquity. On each beach, over eons and eons, God created new life, again and again. To those

[166]   CHAPTER SEVEN

who opened their imaginations to the possibilities of deep time, the meticulous geology of Heade's *Lake George* offered the experience of multiple, simultaneous temporalities. First, the viewer would apprehend the painting in a single, unbroken instant, as James had argued with his stream of consciousness. A second temporality would be suggested by the small boats gliding through the lake, and the clouds drifting across the sky: time would pass slowly, but still comprehensibly as it did in daily life. Finally, the deep time in the rock-portraits of the foreground offered a third temporality, a vision of an earthly past so remote that it defied human comprehension and touched the divine eternity itself.[52]

Conclusion

In *The Book of Sand* (1975), the Argentinian writer Jorge Luis Borges tells the story of a man who buys an infinitely long book. At first the man is fascinated as the pages multiply between the covers. But finally he is horrified that he can find no beginning and no end. "If space is infinite," he despairs, "we may be at any point in space. If time is infinite, we may be at any point in time."[53] Declaring the book monstrous, he hides it in the national library.

We might say that the book of sand was first written in the nineteenth century. Released from the constraints of the biblical chronology, the earth's past now stretched into a time so remote as to be incomprehensible. Here was a world without knowable beginning, without prospect of an end. But instead of hiding from deep time's existential horrors, the scientists, artists, and others who attended its birth stepped forward, striving to understand the thing that they did not understand. In the process they created a haunting array of images that speak to the urgency of their desire to comprehend the incomprehensible. When comprehension ended, they summoned their imaginations. With ink and paint, they crafted beautiful paper worlds filled with curious creatures and strange viewpoints, many rendered in glorious color. Their children are alive and well and living with us today, a recent illustration of geological time shows (fig. 7.9).

We still have a hard time grasping deep time. John McPhee, who coined the term in 1981, said that deep time awed the imagination "to the point of paralysis."[54] But deep time also fills us with wonder, and we too mobilize our imaginations when comprehension ends. That the nineteenth century bequeathed to us a dilemma rather than a solution is therefore perhaps cause for celebration rather than despair.

Notes

1. On the adoption of geological deep time in ethnology and linguistics, see Thomas R. Trautmann, "The Revolution in Ethnological Time," *Man*, n.s. 27, no. 2 (June 1992): 379–97. There is a large literature on deep time in European thought more broadly; major interpretations include: Paolo Rossi, *The Dark Abyss of Time: The History of the Earth and the History of Nations from Hooke to Vico* (Chicago: University of Chicago Press, 1984); Stephen J. Gould, *Time's Arrow, Time's Cycle: Myth and Metaphor in the Discovery of Geological Time* (Cambridge, MA: Harvard University Press, 1987); Martin Rudwick, *Bursting the Limits of Time: The Reconstruction of Geohistory in the Age of Revolution* (Chicago: University of Chicago Press, 2005); and

Rudwick, *Worlds Before Adam: The Reconstruction of Geohistory in the Age of Reform* (Chicago: University of Chicago Press, 2008). For a recent book on the importance of thinking with deep time, see Marcia Bjornerud, *Timefulness: How Thinking Like a Geologist Can Help Save the World* (Princeton, NJ: Princeton University Press, 2018).

2. David Dale Owen, "Scientific Pursuits: Introduction to Geology," *Quarterly Journal and Review* 1 (Jan. 1846): 44; and Jacob Green, *The Inferior Surface of the Trilobite Discovered: Illustrated with Coloured Models* (Philadelphia: Judah Dobson, 1839), 10.

3. Herbert Spencer, *Essays: Scientific, Political, and Speculative*, 3 vols. (1858; repr. New York: D. Appleton and Co., 1892), vol. 1, 230; Charles Darwin, *On the Origin of Species by Means of Natural Selection*, 3rd ed. (London: J. Murray, 1859), 282; and Charles Lyell, *Travels in North America in the Years 1841–42*, 2 vols. (New York: Wiley and Putnam, 1845), vol. 1, 42.

4. Jean-Jacques Rousseau, *A Discourse on Inequality*, ed. Maurice Cranston (1755; New York: Penguin, 1985), 68.

5. On premodern ideas about time (cyclical, multivalent, discontinuous), see Keith Moxey, *Visual Time: The Image in History* (Durham, NC: Duke University Press, 2013); and Denis Feeney, *Caesar's Calendar: Ancient Time and the Beginnings of History* (Berkeley: University of California Press, 2008).

6. James Elkins, "Art History and Images That Are Not Art," *Art Bulletin* 77, no. 4 (Dec. 1995): 553–71; Dennis Cosgrove, "Maps, Mapping, Modernity: Art and Cartography in the Twentieth Century," *Imago Mundi* 57, no. 1 (2005): 35–54; Ann Shelby Blum, *Picturing Nature: American Nineteenth-Century Zoological Illustration* (Princeton, NJ: Princeton University Press, 1993), 3–19; and Donald A. Schon, *Displacement of Concepts* (1959; repr. London: Tavistock Publications, 1963).

7. Chauncey Wright, "Evolution of Self-Consciousness," *North American Review* 116, no. 239 (April 1873): 247: "The forces and laws of molecular physics are similarly related to actual human intelligence. Sub-sensible properties and powers can only be empirically known, though they are 'visualized' in the *hypotheses* of molecular movements and forces."

8. All biographical details about David Dale Owen are from Walter Brookfield Hendrickson, *David Dale Owen: Pioneer Geologist of the Middle West* (Indianapolis: Indiana Historical Bureau, 1943).

9. William Smith, *Strata Identified by Organized Fossils, Containing Prints on Colored Paper of the Most Characteristic Specimens in Each Stratum* (London: W. Arding, 1816).

10. Charles Lyell, *Principles of Geology, Being an Attempt to Explain the Former Changes of the Earth's Surface, by Reference to Causes Now in Operation*, 3 vols. (London: John Murray, 1830–33), vol. 1, 89.

11. William Maclure, "Observations on the Geology of the United States, Explanatory of a Geological Map," *Transactions of the American Philosophical Society* 6 (1809): 427.

12. This is to distinguish these new maps from older maps that pointed out ancient features on an essentially "modern" cartographic plane. See for example the "antique sculptures" (Indian mounds) in Lewis Evans, *A Map of Pensilvania, New-Jersey, New-York, and the Three Delaware Counties; by Lewis Evans* (n.p., 1749).

13. Edwin James and Stephen Long, "Vertical Section on the Parallel of Latitude 41 Degrees North," in *Account of An Expedition from Pittsburgh to the Rocky Mountains, Performed in the Years 1819 and 1820 . . . Under the Command of Major Stephen H. Long* (Philadelphia: H. C. Carey and I. Lea, 1822), David Rumsey Map Collection, David Rumsey Map Center, Stanford Libraries.

14. Owen, "Scientific Pursuits," 40–46.

15. David Dale Owen, Joseph Leidy, J. G. Norwood, C. C. Parry, Henry Pratten, B. F. Shumard, and Charles Whittlesey, *Report of a Geological Survey of Wisconsin, Iowa, and Minnesota: And Incidentally of a Portion of Nebraska Territory* (Philadelphia: Lippincott, Grambo, and Co., 1852), 94.

16. Owen, "Scientific Pursuits," 40–46.

17. Emma Willard, *Willard's Historic Guide. Guide to the Temple of Time; and Universal History, for Schools* (New York: A. S. Barnes & Co., 1849), 15, 12.

18. Willard, *Willard's Historic Guide*, 17.

19. David Dale Owen, "Abstract of the Proceedings of the Fourth Session of the Association of American Geologists

and Naturalists," *American Journal of Science and Arts* 45, no. 1 (1843): 136, 137.

20. "The appearance and outline of a portion of the south shore of Lake Superior, is accurately delineated in the following cut, after a sketch executed, at request, by Aindi-bi-tunk, a full-blooded Chippewa Indian." In Owen et al., *Report of a Geological Survey*, xxxiii.

21. Owen, "Abstract of the Proceedings," 137.

22. Lyell, *Principles of Geology*, vol. 1, 191.

23. Owen et al., *Report of a Geological Survey*, xiii.

24. Owen et al., *Report of a Geological Survey*, 94.

25. Owen et al., *Report of a Geological Survey*, 205.

26. Owen et al., *Report of a Geological Survey*, 50.

27. Lyell, *Principles of Geology*, vol. 1, 8, 10, 82.

28. Herbert Spencer, *The Principles of Psychology* (London: Longman, Brown, Green, and Longmans, 1855), 4. For more on brain science in the nineteenth century, see (in a large literature) Edwin Clarke and L. S. Jacyna, *Nineteenth-Century Origins of Neuroscientific Concepts* (Berkeley: University of California Press, 1987); and Robert Richards, *Darwin and the Emergence of Evolutionary Theories of Mind and Behavior* (Chicago: University of Chicago Press, 1987).

29. Spencer, *Principles of Psychology*, 231, 246.

30. Immanuel Kant, *Critique of Pure Reason*, ed. and trans. Paul Guyer and Allen Wood (Cambridge: Cambridge University Press, 1999), 7–8; and Stephan Körner, *Kant* (1955; repr. New Haven, CT: Yale University Press, 1982), 34, 36.

31. Spencer, *Principles of Psychology*, 52–65.

32. Spencer, *Principles of Psychology*, 246.

33. J. Hughlings Jackson, "Remarks on Evolution and Dissolution of the Nervous System," *Journal of Mental Science* 33, no. 141 (April 1887): 29.

34. Jackson, "Remarks on Evolution," 31, 32; J. Hughlings Jackson, "On Epilepsy and on the After Effects of Epileptic Discharges," in *The West Riding Lunatic Asylum Medical Reports*, ed. J. Crichton Browne and Herbert C. Major, vol. 6 (London: Smith, Elder, and Co., 1876), 267.

35. Chauncey Wright, "Evolution of Self-Consciousness," *North American Review* 116, no. 239 (April 1873): 247.

36. William James, *The Principles of Psychology*, 2 vols. (New York: Henry Holt, 1890), vol. 1, 6, 81, 30.

37. George John Romanes, *Mental Evolution in Man: Origin of Human Faculty* (London: Kegan Paul, Trench & Co., 1888), 2.

38. James, *Principles of Psychology*, vol. 1, 224. The phrase "stream of consciousness" first appeared in an 1884 essay in which James criticized David Hume's view of the mind as an agglutination in various shapes of separate entities called ideas. See Stephen Kern, *The Culture of Time and Space, 1880–1918* (1983; repr. Cambridge, MA: Harvard University Press, 2003), 24.

39. James, *Principles of Psychology*, vol. 1, 279, 280.

40. Roderick Impey Murchison, *The Silurian System*, 2 vols. (London: John Murray, 1839), vol. 1, 6.

41. James, *Principles of Psychology*, vol. 1, 282.

42. Louis Agassiz, "America the Old World," in *Geological Sketches* (Boston: Ticknor and Fields, 1866), 1.

43. Henry T. Tuckerman, *Book of the Artists: American Artist Life* (New York: Putnam, 1867), 557. See also Rebecca Bedell, *The Anatomy of Nature: Geology and American Landscape Painting, 1825–1875* (Princeton, NJ: Princeton University Press, 2001); and Angela Miller, *The Empire of the Eye: Landscape Representation and American Cultural Politics, 1825–1875* (Ithaca, NY: Cornell University Press, 1993).

44. Gotthold Ephraim Lessing, *Laocoön. An Essay upon the Limits of Painting and Poetry. With Remarks Illustrative of Various Points in the History of Ancient Art*, trans. Ellen Frothingham (1776; Boston: Roberts Brothers, 1887), 109.

45. Kern, *Culture of Time and Space*, 20–22.

46. Theodore E. Stebbins Jr., with the assistance of Janet L. Comey and Karen E. Quinn, *The Life and Work of Martin Johnson Heade: A Critical Analysis and Catalogue Raisonné* (New Haven, CT: Yale University Press, 2000), 37–39; and Theodore E. Stebbins Jr., *The Life and Works of Martin Johnson Heade* (New Haven, CT: Yale University Press, 1975), 36–41.

47. Louis Agassiz, "The Glacial Theory and Its Recent Progress," *Edinburgh New Philosophical Journal* 33 (April–Oct. 1842): 217–83.

48. James Geikie, *The Great Ice Age, and Its Relation to the*

Antiquity of Man (New York: D. Appleton and Co., 1874), 381.

49. Louis Agassiz, "The Silurian Beach," in *Geological Sketches* (Boston: Ticknor and Fields, 1866), 29–63.

50. In homage to his friend Adam Sedgwick, who had named the "Cambrian" (the Latin word for Wales) geological stage, Murchison named this slightly younger formation after the Silures, a local Celtic tribe that had resisted the Roman invasion. "Having discovered that the Region formerly inhabited by the Silures, celebrated in our annals for the defence of the great Caractacus, contained a vast and regular succession of undescribed deposits of a remote age, I have named them the 'Silurian System'" (Murchison, *The Silurian System*, vol. 1, v).

51. Agassiz, "Silurian Beach," 37, 31.

52. Other rock-paintings from the 1860s were viewed with the long chronology in mind. A critic reviewing William Stanley Haseltine's paintings of coastal Nahant in 1864 wrote: "Agassiz pronounces the rocks of Nahant to be the oldest on the globe, and that they are of volcanic origin. Mr. Haseltine fully conveys their character in his pictures, and no one who has wandered over those huge masses of rough red rock, and watched the waves breaking against them, could fail to locate, from his studies, the very spot he has delineated." *Watson's Weekly Art Journal*, quoted in Bedell, *Anatomy of Nature*, 121.

53. Jorge Luis Borges, "The Book of Sand," trans. Norman Thomas de Giovanni, *New Yorker* (Oct. 25, 1976): 38–39.

54. John McPhee, *Basin and Range* (New York: Farrar, Straus & Giroux, 1981), 21.

8

Susan Schulten

How Place Became Process
The Origins of Time Mapping in the United States

The new republic of the United States was established in 1776 and dedicated above all else to the idea of a better future. Yet national consolidation and expansion in fact involved a sustained effort to map the past. Over the course of the nineteenth and early twentieth centuries, a series of maps crafted a new story of the United States that unfolded on an ever-widening territorial canvas emptied of native peoples and filled by the relentless forces of "progress." Eventually, these maps began to experiment with causality itself, suggesting that large-scale processes unfolding in space were as significant as more traditional historical subjects, such as battles and treaties. This new cartographic thinking coalesced in the *Atlas of the Historical Geography of the United States* (1932), the most ambitious effort during the twentieth century to map the American past. This essay charts the evolution of temporal thinking that led to this extraordinary compilation, changes that anticipate the rise of Historical GIS in our own time.

Thirty years in the making, the *Atlas of the Historical Geography of the United States* was the brainchild of one of the country's first professionally trained historians. At the turn of the century, John Franklin Jameson (1859–1937) proposed a two-part volume that would both answer existing questions about American history and prompt new ones. It would begin with a selection of archival maps—over a hundred images, reproduced in facsimile—ranging from fifteenth-century portolan charts to nineteenth-century western surveys. But the heart of the atlas would be the more substantial second part, made up of five hundred original maps designed to explore American political, social, economic, and territorial development. One series of maps would trace the distribution of financial capital from 1800 to 1928; another would calculate shifting patterns of land value from the antebellum era to the Great Depression. All were designed to demonstrate the power of cartography to stimulate and support the analysis not just of historical events, but also of *processes*.

The sheer scope and scale of Jameson's project marked it as a breakthrough publication—the culmination of sustained research in archives that had only recently been established, by scholars in disciplines that had only recently been organized. (The first editor of the atlas was another professional historian, Charles Oscar Paullin, who labored for years to realize Jameson's vision; John Kirkland Wright, one of the country's first academic geographers, was

brought in to help complete the project.) All of this placed the *Atlas* at the vanguard of social-science scholarship. But it was the editors' intellectual ambition that distinguished their project. In a particularly striking comment, Wright mused that "the ideal historical atlas might well be a collection of motion-picture maps, if these could be displayed on the pages of a book without the paraphernalia of projector, reel, and screen."[1]

It would be decades before Wright's vision would fully come to pass; only in the twenty-first century are motion-picture or animated maps emerging as a feasible technique for digital humanists. Yet his prophetic comment also reminds us that the effort to transform paper maps into instruments of time was nothing new. In fact, as this chapter will show, the effort to equip maps to capture both time and motion extends back to the early republic.

The Post Office Maps the Mail

The earliest effort to map time in the United States grew from the practical needs of governance. The Postal Service Act of 1792 had vastly expanded the operation and scope of domestic mail service, authorizing new post offices in the remote frontier and acknowledging mail delivery as a federal obligation. The Act transformed the Post Office into a hub of activity and made postal workers the most visible representatives of the new national government. The expanded scope of the Post Office prompted one of its chief clerks, Abraham Bradley Jr. (1767–1838), to map the entire mail network in 1796. The result was the most detailed map of the nation to date, and one that underscored time and movement. Bradley's goal is evident from his selection of information: while most contemporary mapmakers highlighted boundaries and topographical features, he foregrounded communication. His image showed a trunk line along the Atlantic coast, with branches extending into western New York, Maine, Vermont, and the southern frontier. By emphasizing how settlements were connected by the post, Bradley's map indirectly conveyed something important about the lived experience of space as well. Densely settled regions were crowded with nodes and routes, suggesting frequent and voluminous interaction, while more sparsely settled areas showed fewer of each.

To explain how information flowed through this system, Bradley inserted an elaborate chart into the lower right quadrant of the map (fig. 8.1). Through a series of serpentine lines, this inset traced the summer and winter delivery times for mail traveling between northern Maine and southern Georgia, allowing readers to understand the regular rhythms of the southbound and northbound routes in each season. By triangulating this chart with the larger map, readers could grasp the overall pace of the post over any given stretch of the main routes in the country. Such a chart rested on the observation and synthesis of mail delivery data, visually integrating space and time. By publishing such a chart, Bradley acknowledged one of the federal government's earliest responsibilities: the commitment to regularly scheduled delivery of information as a means of unifying the fledgling republic.[2]

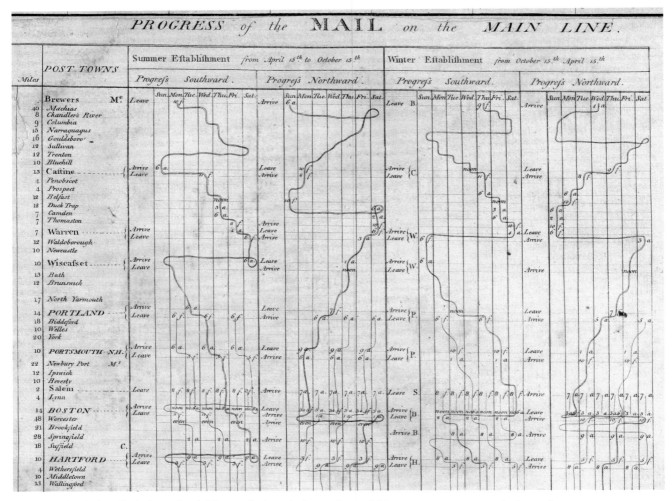

FIG. 8.1 A detail of Bradley's chart of mail delivery, which worked in tandem with his map to trace movement in time and space. From "A Map of the United States Exhibiting Post Roads & Distances by Abraham Bradley, Junr." (1796?). 94 x 88 cm. Courtesy of the Library of Congress.

Emma Willard Maps the Pageant of National History

The expanded commitment to mail delivery in the 1790s answered a critical need to unify the republic in the wake of national independence. Equally essential was the concurrent effort to cultivate civic and national identity. By the 1780s, Noah Webster had begun to assert a uniquely "American" form of English, while Jedidiah Morse insisted that the nation's youth learn geography from countrymen such as himself.

And just as independence influenced the study of language and geography, it also shifted the meaning of history. In the aftermath of the Revolution, educators began to reframe the past as prelude to the recent history of the new United States.

Among the strongest advocates for this new American history was Emma Hart Willard (1787–1870), a visionary teacher who came of age in an era of rapidly expanding educational opportunities for women.[3] Willard's commitment to instilling history and geography through memorization led her to ex-

How Place Became Process [173]

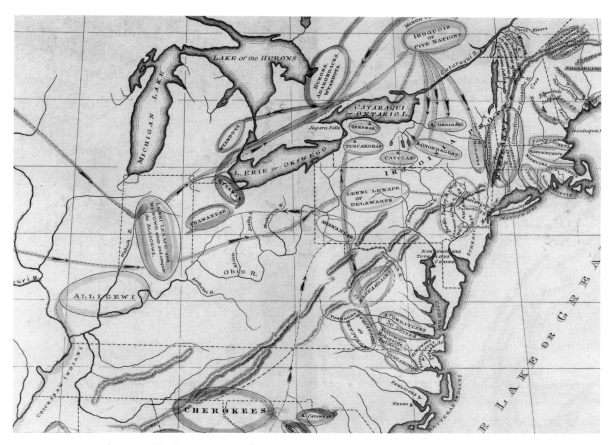

FIG. 8.2 A portion of Emma Willard's 1828 map of indigenous migrations. In contrast to her methodical attention to change over time when depicting the history of Eurasian civilizations, here she collapsed several centuries onto a single plane. From *A Series of Maps to Willard's History of the United States, or, Republic of America* (New York: White, Gallaher & White, 1829). 28 x 31 cm. Courtesy of the David Rumsey Map Collection, David Rumsey Map Center, Stanford Libraries.

periment with visual mnemonic devices, for in her mind the visual preceded the verbal. History literally took *place*; if students could situate past events in geography, they would be more likely to retain that knowledge. As a result, Willard's textbooks were studded with graphic tools, especially maps. For instance, in the early 1820s she contributed a time-map of the ancient world to William Woodbridge's universal geography. To assist memorization, Willard drew the history of ancient Rome in the shape of a mighty river, with conquered civilizations shown as tributaries coming in at various points. Although fanciful, hers was a creative effort to map time. This river motif was also used to metaphorically map the abolition of the slave trade. For both Willard and the abolitionists, the river motif captured a belief that time unfolded progressively, just as social reform ideally would.[4]

Willard then turned her attention from the ancient to the American past in her *History of the United States* (1828). Together with a former student, she designed a separately published "Series of Maps" to accompany the text, creating not just the first published school history of the United States, but also the nation's first historical atlas. After several reprints, these separately published maps were simplified and

incorporated into later editions of the textbook. Willard designed her maps to tell the story of the United States; each successive image depicted either a contribution to or a legacy of nationhood, translating the chaos of the past into a systematic trajectory whose inflection point was the American Revolution. The result was an atlas that framed the United States not just as an idea, but as a territorial progression. In the pages of Willard's atlas, American history unfolded across both space and time, culminating in independence and national fulfillment.[5]

Willard's view of history as the progressive realization of the nation was embedded in the very structure of her atlas, which opened with a map of "aboriginal wandering" across the eastern half of North America (fig. 8.2). Directional arrows conveyed some general migration patterns among the larger native nations, condensing centuries of movement into a single compact image. Tellingly, Willard distinguished this "introductory map" from the numbered series that followed, revealing her widely shared belief that Native Americans existed in a category prior to national history. The anachronistic addition of state borders (marked by dotted lines) served to remind students that native lands would eventually be claimed, owned, and inhabited by an entirely different society.[6]

The same point was reinforced in the written text; native life was discussed in a preface, while the first numbered chapter opened with the arrival of Christopher Columbus. The map that Willard created to illustrate this episode in American history featured prominent European voyages of discovery and settlement in the sixteenth century. Like the introductory image, this "first" map collapsed an extended era onto a single page, albeit preserving a more precise temporal sequence by dating each major voyage and noting the competing European claims to North America. Willard also revealed her vision of history by including failed colonization efforts—notably Humphrey Gilbert's 1578 patent, the earliest English claim in North America—insofar as they anticipated Anglo-American settlement. In this way, even as Willard identified Dutch and Spanish voyages, she implied an inevitably English future. Conversely, as the European penetration of the interior came to dominate subsequent maps, signs of native occupation were gradually removed. In this way, her atlas presented colonial geography from the standpoint of the eventual victors, framing the earliest phases of American life through her own ancestors' eyes. As the proud descendant of English settlers—what she termed "the best of the old puritan blood"—this influential educator effectively mapped American history as the story of her own ethnic group's inheritance.[7]

The success of her historical atlas prompted Willard to develop other graphic devices to help students apprehend—and remember—the interdependence of history and geography. In 1835, she designed a remarkable "Picture of Nations" to illustrate all of human history on a single page. Timelines and time charts had proliferated since the mid-eighteenth century, and depictions of time as a river had appeared in both Europe and the United States by the early 1800s.[8] Where Willard stood out was in the dimensionality and fluidity of her imagery. Drawn in

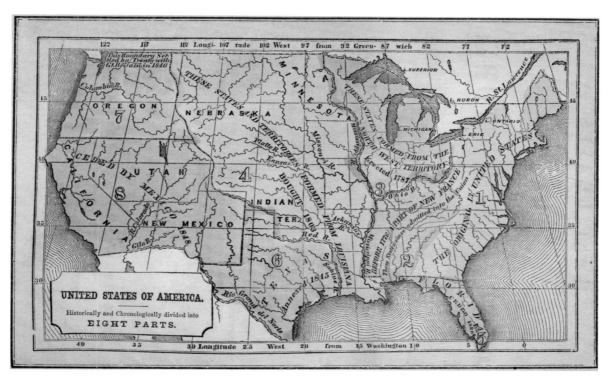

FIG. 8.3 Willard was among the first to frame the nation's territorial acquisitions as seemingly natural steps in a progressive history. From Willard, *Last Leaves of American History: Comprising a Separate History of California* (New York: A. S. Barnes & Co., 1853). 11 × 17.5 cm.

three-point perspective, her visualization of time progressed toward the reader in an ever-widening flow, leaving ample space in the foreground for the modern era (and implying that recent history was more consequential for her readers' lives). A hybrid of map and chronicle, this dramatic image made the past a panorama, elevating students to a remote, god-like position from which they could watch the human drama unfold down the centuries.

In the 1840s Willard similarly fashioned "Temples of Time" to teach both world and American history. This classically inspired tool invited students to plot the past into a grid-like, three-dimensional structure. Pillars along the side walls signposted the march of time from the back to the front of the image (each pillar representing a century), while the breadth of the temple loosely represented geographic space. On the ceiling, students were to plot individuals in their proper chronological spot. Willard intended her Temple to integrate space and time in the same way that a map coordinates longitude and latitude. As she envisioned it, the ceiling would be filled "with the names of those great men who are to history, as cities are to geography." The floor would be likewise inscribed with events that underscored progress, with increased attention to the recent past as time moved forward toward the reader. In her explanation of the Temple of Time, Willard argued that maps were essentially a graphic language, insisting that "it is as scientific and intelligible, to represent time by space, as it is to represent space by space." Several years later she confidently claimed, "In history I have invented the map."[9]

As Emma Willard found herself living through seismic events in the later antebellum era, she raced to incorporate the latest news into her histories. She

commented more than once that time itself seemed to have quickened, a product of advances in transportation technology as well as the nation's rapid westward expansion. These developments fundamentally changed her apprehension of time.[10] The constant need to update her American history text created a compelling context for thinking about the past as bearing on the present. In 1849, she published *Last Leaves of American History*, an account of the war against Mexico and the astonishing territorial growth brought through the Treaty of Guadalupe Hidalgo. In the 1853 edition of that text, Willard proudly incorporated a schematic view of continental expansion, divided into "eight chronological parts" that stretched from the original thirteen colonies to the far western acquisitions of 1848 (fig. 8.3). The map may seem ordinary to contemporary readers, accustomed as we are to its visual program. But Willard was among the first Americans to map the nation's territorial expansion in this way, casting a convulsive and violent history as an ordered process of cessions, acquisitions, and purchases. This benign model of American territorial growth remains a fixture of high school history texts to this day, creating a prime example of what William Rankin (in this volume) terms "collapsed animation."

Johann Georg Kohl Maps the Progress of Geographical Knowledge

At the height of her influence as an educator, Emma Willard toured Europe to attend the World's Educational Convention of 1853. While in London she happened to meet Johann Georg Kohl (1808–1878), a German scholar of North American exploration who was about to embark for the United States. To advance his research, Kohl had spent years painstakingly copying original maps of discovery that he found in archives across Europe.

This kind of archival cartography represented a new approach for Willard, who was impressed by Kohl's erudition and collection. While she had evidently not thought much about old maps until then, Kohl quickly convinced her of their importance. At a time when most Americans valued maps only to the degree that they were accurate, Kohl argued otherwise. "Outdated" maps should not be discarded but rather treated as valuable historical evidence: grist for chronicling territorial growth, shifts in sovereignty, and the accumulation of geographical knowledge. In fact, when he met Willard, Kohl was preparing to bring more than a thousand map manuscripts to the United States to convince the nation's leaders that cartographic materials ought to be preserved, archived, and appreciated alongside written texts as foundational documents of the country's history. From Europe, Willard promptly wrote a letter to the Secretary of State to endorse that endeavor.

Johann Kohl's visit to the United States would eventually influence the practice of map collecting. In an extended address to the Smithsonian Institution in 1856, he revived a dormant proposal from the American Association for the Advancement of Science to build a federal geographical library. Though such a library would not be organized until the end of the century, Kohl was influential in other ways. With

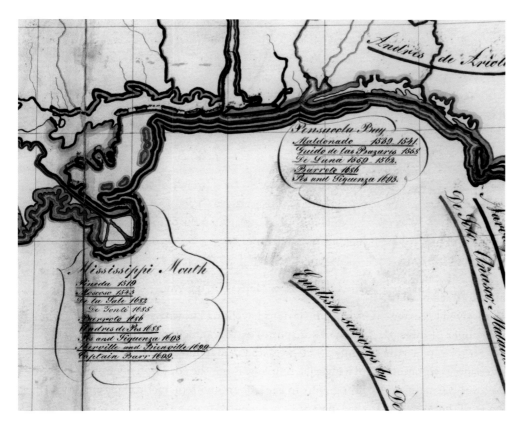

FIG. 8.4 Johann Georg Kohl overlaid successive explorations to show both the progress of discovery and the contingent nature of knowledge. Kohl, "A Map Showing the Progress of the Discovery of the Gulf of Mexico" (1856). 49 x 87 cm. Courtesy of the Library of Congress.

enthusiastic support from Willard, among others, the State Department hired Kohl to copy 474 of his facsimile maps relating to North American exploration. The resulting collection of copies seeded a new Hall of Charts and Maps—the core of what would later become the Geography and Map Division at the Library of Congress. Although the collection remained relatively neglected for decades, it would be discovered and extensively mined at the turn of the twentieth century by Justin Winsor (exactly as Kohl had hoped) as a resource for reconstructing early American history.[11]

While Kohl was busy retracing his facsimile maps for the State Department, he also conducted cartographic research for the US Coast Survey in 1856 and 1857. The Coast Survey had a mandate to chart the nation's coastlines and waterways, which meant that the western lands acquired in the Treaty of Guadalupe Hidalgo in 1848 significantly expanded its scope of responsibility. Even as Director Alexander Bache assigned several of his men to map the nation's new Pacific Coast, he commissioned Kohl to compile its history. While it was unusual at the time for a federal agency to devote precious resources to mapping the past, Bache had a strong sense of his agency's role in history. The Coast Survey had begun to build its own archive, and Kohl's attention to the "progress of discovery" on the Pacific Coast was integral to this broader effort. The Survey might be primarily focused on charting the waterways and coasts in the present, but it was also attuned to the accumulation of geographical knowledge in the past.

Pleased with Kohl's completed history of discovery along the Pacific Coast, Bache commissioned him to assemble similar narratives for the Gulf and Atlantic coasts. For each of these reports Kohl cre-

ated a distinctive map of exploration, overlaying different colored bands to show the extent of successive voyages along a given coastline over time (fig. 8.4).[12] By schematically highlighting the chronology and scope of each noted expedition, these innovative maps created a visual genealogy of European discovery. In the interior, Kohl removed all reference to native life (an interesting omission, given his extensive work with native tribes in the upper Midwest). Instead, these parts of each map were inscribed with highlights from the history of overland exploration. Thus, the Pacific Coast map juxtaposed the maritime voyages of Francis Drake in 1579 with the routes of early European explorers in the interior West.

Taken together, Kohl's life-work extended the larger nineteenth-century drive to collect, archive, and organize information into the domain of cartography. His archive would not only enable later historians to undertake cartographic research but would influence the way Americans viewed old maps. Gouverneur Kemble Warren would acknowledge Kohl's model in his own landmark 1858 map of the trans-Mississippi West; likewise, George Wheeler would later confirm Kohl's influence over his ambitious western survey that began in 1869. Inspired by Kohl, Wheeler reproduced historic maps of the west in his published report, placing them alongside the latest and most geographically accurate information.

As these examples confirm, the German historian Johann Kohl exposed American archivists, surveyors, and mapmakers alike to a new method of cartographic inquiry. Valuing maps not only for their geographical accuracy but also for the stories they could tell about the progress (and limits) of geographical knowledge in the past, he effectively introduced Americans to a long-standing European method of making maps tell time.[13]

The Coast Survey Maps the Civil War

As map historians have long known, the Civil War was a major spur to cartographic innovation in the United States—with talented immigrants like Kohl playing a leading role. From the adoption of photolithography to experiments with thematic mapping, the Coast Survey benefited from an influx of skilled Germans in the 1850s. When Superintendent Bache fell ill in the spring of 1862, it was German Julius Hilgard (1825–1890) who oversaw the agency for the duration of the Civil War. Shortly after stepping in as acting superintendent, Hilgard ordered the creation of a recent-events map to boost Union support for the war. Drawn by another immigrant, Henry Lindenkohl, this "Historical Sketch of the Rebellion" highlighted the military progress of the Union up to the twin victories at Forts Henry and Donelson and the capture of New Orleans. Just as Kohl drew colored lines to mark successive voyages, Lindenkohl used a similar technique to dramatize recent Union advances.

Eager to highlight success, the Survey regularly revised and reissued this map to mark the progress of the war. Updates came not at set intervals but after significant gains. In July 1863, for instance, the survey published three separate editions to highlight the victories at Gettysburg and Vicksburg.[14] At a time when

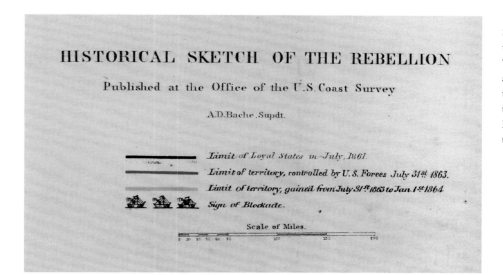
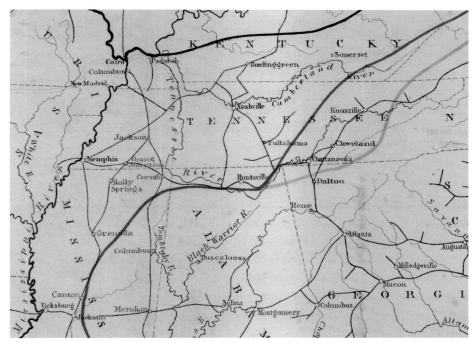

FIGS. 8.5A AND 8.5B Details from the 1864 edition of the Coast Survey's "Historical Sketch of the Rebellion," a series of maps designed to publicize territorial gains in the war against the Confederacy. Drawn by Henry Lindenkohl. 47 x 45 cm. Courtesy of the Library of Congress.

rioters in New York City were expressing fierce opposition to the recent draft and frustration with the Lincoln administration's conduct of the war, these news-maps carried more than a little urgency. Ideally, a cartographic demonstration of the Union's latest successes would secure public support. Lindenkohl's maps had their work cut out for them.

A closer look at the resulting "Historical Sketch" maps shows the logic behind their form. While the prototype of April 1862 included individual battles and commanders, later editions presented Union success in more general terms. In other words, the editors decided not to map discrete victories but to extrapolate (and maximize) the territorial advanc-

es brought by these victories. Rather than focusing on battles, each updated "Historical Sketch" painted a larger, more generalized, and more positive picture of the Union effort. Each edition also reveals an effort to reframe the struggle. At first, Lindenkohl drew just two lines (red and blue), contrasting Union control of territory at the war's outset in July 1861 with its holdings as of spring 1862. But in early 1864 (figs. 8.5a and 8.5b), he added a third, yellow line to highlight the substantial victories at Gettysburg and Vicksburg, as well as recent gains in Tennessee, along the Red River, and along the Texas coast. The shifting fortunes of the war in the present were prompting Lindenkohl to reorganize the time frame of the past.

Lindenkohl's maps must have struck a chord, for similar visualizations cropped up in other contemporary depictions of the war. At a moment of severe crisis on the home front in the summer of 1863—when an antiwar faction threatened to undermine the Lincoln administration by forging a separate peace with the Confederacy—a committee of fervent Unionists issued a broadside to argue for staying the course. An accompanying map adapted Lindenkohl's technique to favorably contrast the extent of territory controlled by the Union in 1861 and 1863. The image—entitled "The Progress of the Union Armies"—handily sidestepped the inconvenient truths that much of the Confederate population still lived in the areas beyond the Union line, and that territorial "control" was in any case a fraught claim.

The same broad-brush method for mapping Union advances appeared in another defense of the war in 1864. That summer and fall, General William Tecumseh Sherman tore through Tennessee and then southeast into Georgia, capturing Atlanta on September 1. News of Sherman's victories buoyed Union spirits, ending doubts about Lincoln's reelection. At that moment, the Army Corps of Engineers turned to maps to convey the significance of recent military victories to the Senate. Corps Chief Engineer Richard Delafield annotated two identical commercial maps of the country to depict those areas under Union and rebel control in 1861 and 1864, respectively. In the first, he stretched the arena of initial Confederate influence to include the ostensibly loyal states of Missouri and Kentucky. This exaggerated the progress recorded on the second map, which incorporated Sherman's massive gains of 1864. While Virginia might remain rebellious, it was undeniable that the army had made progress. The navy now controlled much of the southern coast, and it had fractured the Confederacy through its victories on the Mississippi River. In fact, the Union advance through Georgia got ahead of Delafield's cartography. Just days after his maps were sent to the Senate, General Sherman alerted President Lincoln that Savannah had been conquered, and the army was poised to push north through the Carolinas.[15]

The Civil War did not drive historical mapping in a vacuum, but it did provide a context, an audience, and an imperative to experiment with maps of time—just as political independence had stimulated the earlier effort to establish a geographical backstory for the new republic. And like that earlier effort, the surge of innovations spurred by the war would

have a lasting legacy. Lindenkohl's model of marking successive stages of Union control in the Civil War is still used in the US Army's official military history, and remains the most common way to visually narrate the progress of the American Civil War to this day.[16] As we will see, it would also leave its mark on the pioneering *Atlas of the Historical Geography of the United States* of 1932.

The Office of the Census Maps the Population

If some of the most thoughtful and innovative efforts to map time during the late eighteenth and early nineteenth centuries were undertaken by the US Post Office and Coast Survey, then the Census Office under the leadership of Francis Amasa Walker (1840–1897) would become the next hub of experimentation. A Civil War veteran of enormous drive, Walker supervised the Ninth Census of 1870 while serving simultaneously as superintendent of the Bureau of Indian Affairs (and later as president of the Massachusetts Institute of Technology).[17] In all three capacities, he acted on his conviction that statistical thinking and mapping were essential to effective administration.

Walker's faith in statistical mapping led him to prepare an unprecedented three-volume *Report of the Ninth Census* (1872). Drawing on new techniques imported from Europe—and considerable immigrant talent in the United States—Walker's *Report* mapped the nation's landscape, climate, and population. The success of this densely cartographic volume prompted him to plan an even more ambitious national statistical atlas, the first of its kind published anywhere in the world. Expanding on the themes in the *Report*, the *Statistical Atlas of the United States* (1874) set a standard that would be widely emulated at home and abroad. While much has been written about this atlas, less attention has been paid to its creative efforts to map change over time.[18] Three examples illustrate the point.

The first is Walker's decision to include a map of the nation's territorial expansion to the Pacific Ocean. Found in both the *Report of the Ninth Census* of 1872 and the *Statistical Atlas* of 1874, this map followed those produced by Willard and others since 1848, showing a string of advances across the continent. But Walker's version was more elaborate, with detailed annotations about each territorial addition. One reviewer welcomed the map as an essential image that ought to hang on the wall of every American schoolroom, for "Nothing can more clearly and impressively epitomize the history of the United States."[19]

The second temporal dimension of the atlas was a more original visualization, involving population data. In the initial *Report of the Census*, Walker and his collaborators experimented with dasymetric (volumetric) mapping in order to show population distribution "as it truly existed." In these maps, a range of shade densities denoted different levels of population density per square mile, with gradients emerging from the data rather than preset by administrative boundaries. With the *Statistical Atlas*, Walker deployed this technique to map not just the 1870 returns but all previous decennial censuses

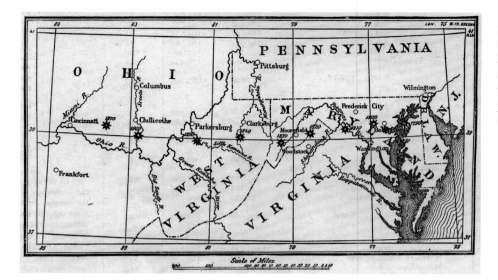

FIG. 8.6 Julius Hilgard's 1872 map of the shifting center of American population was incorporated into Walker's *Statistical Atlas* (1874), and would be widely emulated thereafter. Map from p. 6 of the *Atlas*, measuring 6 x 12 cm. Courtesy of the Library of Congress.

since 1790. The resulting maps were then arranged in a time series to evoke a visual narrative of expanding settlement.

A closer look at this series reveals the historical thinking that animated Walker and his colleagues. In each map, a small star was placed at the statistical "center" of population density, a location that moved progressively westward across Maryland into Virginia, West Virginia, and southern Ohio. Walker was enthusiastic about mapping movement, and he opened the atlas with a small snapshot of that migration in order to signal the themes ahead (fig. 8.6). The map appears to have been the brainchild of Julius Hilgard, the same superintendent of the Coast Survey who supervised the "Sketch of the Rebellion" series during the Civil War.

In an article for *Scribner's Monthly*, Hilgard explained that it was the unique geographic context of American history that had prompted the Census Office to calculate the center of population in the first place. Whereas Europe had long been settled to capacity, the United States had repeatedly opened new lands for migrants. This dynamic created the imperative for observers, armed with past census data, to map and study migration in meaningful and creative ways. This was no idle investigation, Hilgard wrote, but rather the foundation for deeper inquiries about American public life. Individual settlers had left their mark on the land through a century of westward migration, and the census allowed social scientists to investigate the aggregate spatial patterns resulting from myriad personal decisions.[20]

Extrapolating from census data, Julius Hilgard and his colleagues developed a rough estimate of the geographic center of the population at each decade. As he explained, the successive opening of "free" lands to the west had invited settlers to make choices about terrain, climate, and surroundings in deciding where to settle. Hilgard's assumptions about freedom of movement were highly exaggerated—even galling—in light of the circumscribed mobility of immigrants and African Americans, not to mention the forced removal of Native Americans. But his goal of studying movement is what interests us here, for this led him to invest his maps with a dynamic, temporal dimension. By chronicling the nation's shifting center of gravity, Hilgard's moving stars approximated a traveler's journey into the trans-Appalachian West. Here was an early iteration of what Rankin in this volume identifies as the "small multiples" style

of historical mapping. In calling this a study of population "advance," Hilgard invoked both senses of the word: to map the physical advance of the population was also to chart material and social progress.

The third historical element of Walker's *Statistical Atlas* was arguably even more innovative: namely, introducing a colored feature to mark the edge beyond which, as of 1870, the population fell below two persons per square mile. Walker drew this "frontier line" (a phrase he coined) as a continuous boundary that roughly followed the 97th meridian; in figure 8.7, this appears as the thick, undulating blue line on the western edge of the map. That line ended up framing all of Walker's population maps, whatever their thematic foci (from the distribution of immigrant groups to literacy and mortality). It was this continuous line, reproduced again and again in Walker's *Statistical Atlas*, that evidently prompted Frederick Jackson Turner (1861–1932) to formulate his famous thesis of the frontier as the central geographical feature of American life.

Turner had immersed himself in cartography beginning in the 1880s. He pressed statistical maps on his students, even as he used them to advance his own research. He was particularly taken with the 1883 *Statistical Atlas* (based on the Tenth Census), which was modeled on Walker's innovative *Atlas of 1874* (based on the Ninth Census). The influence of these atlases is clearly visible in Turner's historical thinking; the demographer's moving frontier line—embedded in the very structure of Walker's maps—gave the historian a powerful tool to conceive of how migration had influenced other social patterns in American life. Simply put, the frontier thesis was a spatial concept; without maps, it could neither be observed nor explained. Little wonder, then, that Turner sent his seminal essay—"The Significance of the Frontier in American History"—to Walker, who praised it and acknowledged its cartographic roots.

The fruitful marriage of history to cartography that Walker had pioneered would far outlast Turner's particular application of it. By demonstrating that maps could be used as instruments of analysis as well as display, Walker anticipated what is now the flourishing use of maps to explore spatiotemporal patterns.[21] Widely reproduced in contemporary texts, his maps prompted historians to investigate spatial patterns in things like voting behavior and land use. They also inspired historians in the late nineteenth and early twentieth centuries to try their hand at statistical cartography.

One important exemplar of this approach was Albert Bushnell Hart (1854–1943). Hart's long-selling *Epoch Maps Illustrating American History*, emphasizing themes of political and territorial growth, was available to students in five editions published between 1891 and 1917. His map of territorial expansion, modeled on precedents made by Walker and Willard, became a common teaching tool in American schools.[22] In the early twentieth century, it even inspired a "mechanical" version where students used pulleys to recreate the stages of westward settlement. With each pull, the "natural" landscape of indigenous tribes and "empty" land was replaced with a phase of settler society moving west. These before-and-after images reinforced the narrative of American expan-

FIG. 8.7 Francis Amasa Walker's "frontier line" was introduced in the *Statistical Atlas* of 1874. On this map, the thick, dark blue line runs roughly along the 97th meridian, but also marks interior "frontiers" in Northeastern Maine, central New York, and around the Great Lakes. As the legend reads, this line marks "the outside limit of a population of 2 or more to the Square Mile." Courtesy of the Library of Congress.

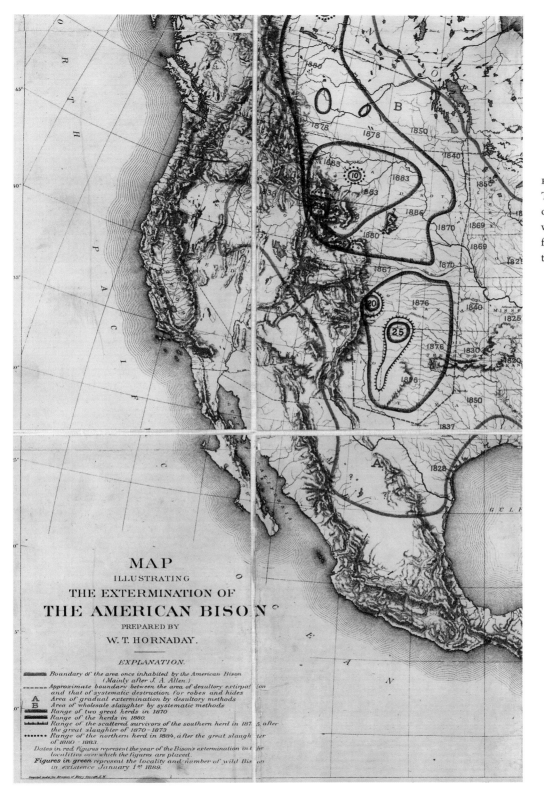

FIG. 8.8 A portion of William Temple Hornaday's 1889 map of the bison, which coincided with Turner's reflections on the frontier. 61 x 58 cm. Courtesy of the Library of Congress.

sion as a series of inevitable waves of migration into open land.²³

But the same techniques could be deployed to depict the dark side of westward expansion as well. To mark the nation's centennial in 1876, historian Joel Allen (1838–1921) turned to cartography to document the near extermination of the North American bison. Starting with a base map that identified rivers and topography, Allen added emigrant routes and railroad lines as well as state boundaries. He then used color to show the contraction of the bison range—as well as its fragmentation by the transcontinental railroad. Allen's map effectively highlighted the ecological cost of a "century of progress." Ten years later, William Temple Hornaday would adapt Allen's visual idiom to amplify the message (fig. 8.8). With successively smaller circles to dramatize their shrinking domain, Hornaday conveyed the decimation of North America's bison over a hundred-year period. Red numbers marked the date of their last sighting in a given area, starkly demonstrating how diminished the herds were from the not-so-long-ago day when they could be found right across the continent, from east of the Allegheny Mountains south to Mexico and north into Canada.²⁴

Hornaday's visualization of environmental catastrophe became an early flashpoint in the conservation movement that flourished under Theodore Roosevelt. It also evoked the simultaneous and similarly violent dispossession of Native Americans. Of these tribes, the northern Sioux were most directly affected by the decline of the bison. With an eighteenth-century domain that ranged from the Rocky Mountains in the west to Minnesota in the east, from the Platte River north to the Yellowstone River, the Sioux had once presided over a bison-rich dominion. Hornaday's map showed the meager remains of the northern herd by 1889, when their confinement to spaces such as the newly named Yellowstone National Park tragically paralleled the forced relocation of the western tribes to reservations. While triumphant maps of westward expansion proliferated through American culture, Hornaday's map showed a similar image in negative. The extermination of the American bison illustrated the underside of Turner's thesis that, by late century, the American frontier had disappeared.

The Atlas of the Historical Geography of the United States

It was in this context that historian John Franklin Jameson envisioned his comprehensive atlas of American history—the monumental work that was briefly described at the start of this essay. Jameson was a classmate of Fredrick Jackson Turner's at Johns Hopkins University (one of the nation's earliest institutions of graduate training), where he studied under historian Herbert Baxter Adams and geographer Daniel Coit Gilman. His exposure to history and geography nurtured his interest in the interdependence of time and space, while also persuading him of the value of old maps as historical documents. Jameson's attention to the history of cartography was further stimulated by Justin Winsor, who relentlessly foregrounded early maps in his *Narrative and Critical History of America*. Finally, Jameson was enthralled

by the new statistical atlases produced by the Census Office in 1874 and 1883. As noted above, by the early twentieth century the enthusiasm for historical mapping—particularly in the statistical and analytical guise modeled by Walker—infused the emerging university disciplines. All of these influences combined to convince John Franklin Jameson that the time was ripe for a comprehensive atlas of the American past.

Jameson initially proposed the project to his former mentor, Daniel Gilman, shortly after Gilman was appointed president of the Carnegie Institution in 1902. The work ought to open, he argued, with a selection of old maps in facsimile: images that would illustrate voyages of discovery while illuminating contemporary perceptions of geography and suggesting early land use. As it eventually took shape under the editorial direction of Charles Paullin (1869–1944), this section of Jameson's pioneering *Atlas of the Historical Geography of the United States* would include over a hundred facsimiles—the largest published collection of historical maps of North America to date. In curating this section, Paullin was guided by Johann Kohl's model to select maps that captured the widening of geographical knowledge over time. It is worth noting that Paullin's collection of facsimiles could not have been assembled in Kohl's day; many of the archives he consulted had only come into existence a few decades earlier. In a pattern that would be repeated across the globe, twentieth-century mapmaking was boosted by nineteenth-century map collecting.[25]

Even more ambitious was the second half of the atlas, for which yet another scholar was brought in—geographer John Kirkland Wright (1891–1969)—to help Paullin document American political, social, and territorial development in a raft of original maps. Some would present what were by then familiar tropes: cartographies of military conflict, native tribes, imperial rivalries, and explorers' routes. Others would chart less familiar phenomena: land grants, agricultural patterns, political parties, and transportation networks. Even where the data and tools used to map these developments were fairly new, however, the impulse behind them echoed the social scientific sensibility that had driven Walker at the Census Office in the 1870s—with one key difference. Whereas Walker had not necessarily made his census maps with an eye to prompting historians to think differently about the frontier, Jameson and his collaborators explicitly approached their project as a way to inspire new historical and geographical questions. Just as digital humanists now design maps as instruments for historical analysis—as noted in this volume's introduction—so too did Jameson hope to advance historical scholarship through maps of time at the turn of the last century.

The 1932 *Atlas* presented a sampler-box of the nineteenth-century experiments with mapping time that have been introduced throughout this essay. To chronicle French exploration in the interior and American expeditions in the far west (plate 39), Paullin and Wright mimicked Johann Kohl's use of colored lines. To render the progress of the Union during the Civil War (plate 163d), they adapted Henry Lindenkohl's "sketch of the rebellion." Their map

of native peoples' migrations (plate 33) was modeled on Albert Gallatin and John Wesley Powell's maps of the same, which in turn showed the influence of Emma Willard's "Aboriginal Wandering." The use of serial maps to chronicle the history of population density, migration, urbanization, economics, elections, and political reform derived from Walker's experiments of 1874. To the degree that the *Atlas* was a scholarly enterprise, it was indebted as much to recent cartographic innovations as it was to recent archival collections. It simply could not have been either conceived or realized even a few decades earlier. In all these respects, the *Atlas of the Historical Geography of the United States* was itself a product of history.

Yet even if their maps clearly depended on newly organized historical archives and emerging disciplinary expertise, Jameson, Paullin, and Wright also broke new ground. For instance, by mapping a succession of elections, they sought to ask and answer new questions about the geography of political power in an expanding democracy; by showing regional disparities in the distribution of reform initiatives over time, they provided new evidence for Fredrick Jackson Turner's concept of sectionalism. Similarly revealing was their effort to map the history of land settlement and cultivation (plates 145–46). As noted above, Wright wrestled with the challenge of graphically representing change over time in his census data, musing that the ideal maps would actually be motion pictures. With no animation techniques at his disposal, he approximated the same effect by developing a series of maps to depict the history of land cultivation (fig. 8.9). Strikingly, instead of showing *all* land under the plow at a given moment in time, each map in the series showed only the latest *additions*. The effect was to draw attention to those regions where land improvements were concentrated in a given decade of American history. Without the benefit of digital tools, Wright had designed a system of data representation that invested maps with a degree of analytical muscle. By crafting a cartographic vocabulary that visually distinguished between areas of more and less rapid change, Wright also anticipated developments in Historical GIS, where sophisticated techniques for mapping *rates* of change are now par for the course.

In short, the quest to map time has a history of its own. Over the course of the nineteenth century, visionary mapmakers directly engaged with graphic and cartographic tools to make sense of America's past in light of an ever-changing present. For Emma Willard and Johann Kohl, the past was framed as a chronicle that led to continent-wide dominion. Neither Kohl nor Willard, however, used maps to engage with causation in a significant way. Instead, their early attempts to map the past were designed primarily to illustrate a known past, marking change rather than inquiring after its spatial and geographical circumstances. In this respect, Kohl's maps of discovery, like Willard's maps of territorial expansion, approximated timelines. Similarly, the Coast Survey's "Historical Sketches of the Rebellion" were designed to mark the Union Army's advances but not to explore the geographical dimension of the war in a sustained manner.

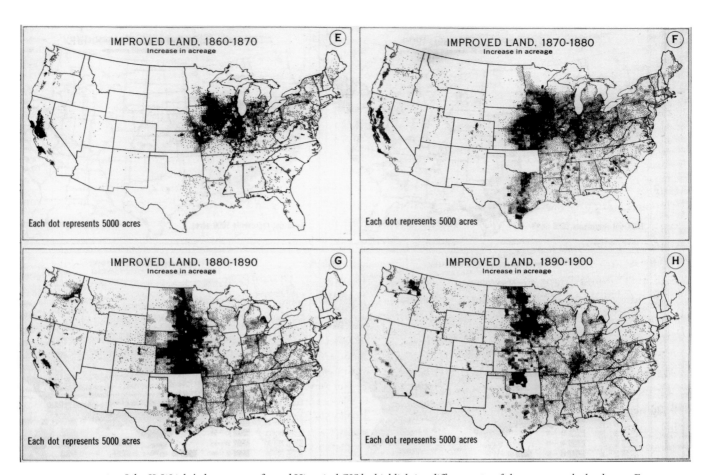

FIG. 8.9 John K. Wright's dot-maps prefigured Historical GIS by highlighting different *rates* of change across the landscape. From Paullin and Wright, *Atlas of the Historical Geography of the United States*, 4 maps on sheet 24 cm wide. Image courtesy of the David Rumsey Map Collection, David Rumsey Map Center, Stanford Libraries; copyright Carnegie Institution of Washington/Carnegie Science.

By the late nineteenth century, the formalization of academic disciplines and federal agencies facilitated a more thorough and systematic engagement with historical mapping. Julius Hilgard's map of the shifting center of population distilled a great deal of census data down into a single, symbolic arc of westward movement. And Francis Walker's efforts to map the distribution of specific population traits—from disease to wealth, race, and ethnicity—drew Frederick Jackson Turner to probe spatial causes as well as spatial effects. Turner's frontier thesis, in turn, inspired Charles Paullin and John Kirkland Wright to design maps that explored more time-space correlations, from the electoral strength of the Republican Party in different counties to the cultivation of wheat on the Great Plains. Maps were both a cause and a consequence of new historical thinking.

It is fitting, then, that the Digital Scholarship Lab (DSL) at the University of Richmond recently sought to bring the *Atlas of Historical Geography of the United States* into the digital realm. The first step in the DSL's "American Panorama" project was to digitize the entire 1932 atlas—an implicit acknowledgment that Jameson, Paullin, and Wright's project was a kind of

ground zero for modern historical spatial analysis. The lab is now pushing further, revisiting the themes of the landmark *Atlas* with new datasets and new digital tools to uncover spatial patterns that would otherwise remain unknown to historians. One project aims to illuminate the unintended consequences of urban renewal by examining patterns of resident displacement. Another investigates the economic implications of industrialization by measuring the distribution of commodities during the canal boom. Although the digital tools at work here are inarguably more powerful than anything Wright could have imagined, the guiding impulse of capturing chronological and chorological change remains strikingly similar. For, as we have seen, Wright's imagination of cartographic motion pictures was but one prophetic moment in what is by now a centuries-long effort to empower maps to measure not just space, but time.

Notes

1. Charles O. Paullin and John K. Wright, *Atlas of the Historical Geography of the United States* (Baltimore: A. Hoen & Co., 1932), xiv.

2. Richard R. John, *Spreading the News: The American Postal System from Franklin to Morse* (Cambridge, MA: Harvard University Press, 1995), 70, 101.

3. Martin Bruckner, *The Geographic Revolution in Early America: Maps, Literacy, and National Identity* (Chapel Hill: University of North Carolina Press, 2006); and Susan Schulten, "Emma Willard and the Graphic Foundations of American History," *Journal of Historical Geography* 33, no. 3 (July 2007): 542–64.

4. Emma Willard's river of history is in *A System of Universal Geography* (Hartford, CT: Oliver D. Cooke & Sons, 1824).

Thomas Clarkson's map of abolition (ca. 1808) can be found at https://brbl-dl.library.yale.edu/vufind/Record/3439900.

5. Emma Willard, *History of the United States, or Republic of America* (New York: White, Gallaher & White, 1828). On the relationship among national identity, territorial sovereignty, historical narratives, and mapping, see Susan Schulten, *Mapping the Nation: History and Cartography in Nineteenth-Century America* (Chicago: University of Chicago Press, 2012).

6. About a decade later, Albert J. Gallatin, former Secretary of the Treasury, created a map of Indian languages not unlike Willard's. Gallatin's map shaped Horatio Hale's map of Indian nations as well as John Wesley Powell's later map of Indian language and settlement. Albert Gallatin, "Map of the Indian Tribes of North America about 1600 A.D."

7. Emma Willard to William Coggswell, Jan. 10, 1842, box 2a, folder 7 of the Education Collection, Sophia Smith Collection and Smith College Archives, Northampton, MA.

8. Willard, "Picture of Nations" (Hartford, CT: F. J. Huntington, 1836), https://www.davidrumsey.com/luna/servlet/s/h40b7z. Willard's perspective sketch was similar to Friedrich Strass's *Strom der Zieten*, as well as Stephen and Daniel Dod's 1807 chart of history. Daniel Rosenberg and Anthony Grafton, *Cartographies of Time: A History of the Timeline* (Princeton, NJ: Princeton Architectural Press, 2010), 143–45; and Schulten, "Emma Willard and the Graphic Foundations of American History," 558.

9. Emma Willard, *Guide to the Temple of Time* (New York: A. S. Barnes & Company, 1850), 21–22; and Willard to Miss Foster, Nov. 5, 1858, reprinted in J. Lord, *The Life of Emma Willard* (New York: D. Appleton and Company 1873), 228.

10. Willard to Jane Hart, March 10, 1848, in *The Papers of Emma Hart Willard, 1787–1870*, Research Collections in Women's Studies, Bethesda, MD, available from the UPA Collection from LexisNexis, 2004, reel 3, frame 280 (hereafter *Willard Papers*); Willard, *Guide*, 4; and Willard to Austin W. Holden, Sept. 5, 1846, *Willard Papers*, reel 3, frame 117.

11. Kohl's attention to facsimile maps paralleled their proliferation in the general culture. The advent of lithography facilitated the growth of facsimile maps of early America, which

became increasingly common tools to access the past. See J. G. Kohl, "Substance of a Lecture Delivered at the Smithsonian Institution on a Collection of the Charts and Maps of America," *Annual Report of the Board of Regents of the Smithsonian Institution* (lecture delivered Dec. 1856) (Washington, DC: Cornelius Wendell, 1857); Justin Winsor, *The Kohl Collection of Maps Relating to America* (Washington, DC: GPO, 1904); and Justin Winsor, *Narrative and Critical History of America*, 8 vols. (Boston: Houghton, Mifflin, and Co., 1884–1889).

12. Like Kohl's 474 facsimile maps of discovery, these reports of coastal exploration are now in the Library of Congress.

13. Gouverneur K. Warren, "Memoir to accompany the map of the territory of the United States from the Mississippi River to the Pacific Ocean, giving a brief account of each of the exploring expeditions since A.D. 1800" [1859], 16. *Report upon United States Geographical surveys west of the one hundredth meridian, in charge of First Lieut. Geo. M. Wheeler . . .* (Washington, DC: Government Printing Office, 1875–1889), v. 1, "Geographical Report," pp. 481–end. As Kären Wigen's essay in this volume notes, Europeans had included old maps in facsimile as early as Ortelius's *Parergon* atlas from the late sixteenth century.

14. The "Historical Sketch" maps were engraved by the German Charles G. Krebs. Six are held at the Library of Congress.

15. "Limit of Territory Controlled by U.S. Forces, January 1861 and Novr. 1864," maps 38(8) and (9), in Record Group 46: Records of the US Senate, National Archives Identifier 305453.

16. Maurice Matloff, *American Military History* (Washington, DC: Office of the Chief of Military History, US Army, GPO, 1969), map 34.

17. Before the Census became a federal bureau, it remained subject to Congressional renewal of its mandate every ten years.

18. *Report of the Ninth Census* (vols. 1 and 2) (Washington, DC: Government Printing Office, 1872); and *Statistical Atlas of the United States Based on the Results of the Ninth Census* ([New York]: Julius Bien, Lith., 1874).

19. "Francis A. Walker's Statistical Atlas of the United States," *North American Review* 121, no. 249 (Oct. 1875): 441.

20. "The Advance of Population in the United States," *Scribner's Monthly* 4 (1872): 214–18.

21. See for example the Spatial History Project at Stanford University, the Digital Scholarship Lab formerly at the University of Virginia and now at the University of Richmond, and the work of Anne Knowles at the University of Maine.

22. Hart's influence ensured that most of the historical mapping at the turn of the century reflected Walker's statistical models. Dixon Ryan Fox, for instance, mapped the history of key elections, Congressional votes, and population shifts. Fox, *Harper's Atlas of American History* (New York and London: Harper & Brothers, 1920). This atlas was made of maps taken from Hart's "American Nation" series.

23. Albert Bushnell Hart, *Epoch Maps Illustrating American History* (New York: Longmans, Green, & Co., 1891). "The Ives Historical Map: A Mechanical Contrivance for Illustrating the Growth of the United States of America," reviewed in *The History Teacher's Magazine* 2, no. 8 (April 1911): 186. By the turn of the century, this magazine included an entire section on "maps" in its roundup of teaching materials, reflecting the proliferation of historical atlases, maps, charts, and other tools.

24. J. A. Allen, *The American Bisons, Living and Extinct* (Cambridge: Cambridge University Press, 1876); and William T. Hornaday, "The Extermination of the American Bison," in *Annual Report of the Board of Regents of the Smithsonian Institution* (Washington, DC: Smithsonian Institution, 1889), map facing p. 548.

25. Jameson's proposal can be found in Ruth Anna Fisher and William Lloyd Fox, eds., *J. Franklin Jameson: A Tribute* (Washington, DC: Catholic University of America Press, 1965), 77–79. See also Jameson to Gilman, Feb. 14, 1902, in *An Historian's World: Selections from the Correspondence of John Franklin Jameson*, ed. Elizabeth Donnan and Leo F. Stock (Philadelphia: American Philosophical Society, 1956).

9

James R. Akerman

Time, Travel, and Mapping the Landscapes of War

Time and Space in Travel Mapping

If all maps tell time, maps made for historical tourists must effectively tell time in several ways at once. On the one hand, their designers are charged with conveying how a complex sequence of events unfolded in space (typically involving the movement of many actors across a given landscape). On the other hand, to the extent that they are meant to be useful to hypothetical visitors, they are simultaneously charged with guiding travelers to and through the landscape, with careful consideration of the rhythm and pace of their movement—whether on foot or horseback, by train, or in an automobile. In abstract terms, the challenge for this kind of map is to do justice to both the event-time of the past and the bodily movement-time of the present. Since the latter is profoundly affected by transportation technology (the means by which travelers physically reach and tour the sites), major technological changes over the past two hundred years have fundamentally altered the ways that historical sites are mapped for tourists.

A pair of guidebooks designed to commemorate the Lewis and Clark Expedition of 1804–1806 make a useful case in point. In 1900, on the eve of the expedition's centennial, the Northern Pacific Railway's annual illustrated companion for travelers on its main line (which ran from St. Paul, Minnesota, to the Pacific Northwest) opened with an extensive article entitled "On the Trail of Lewis and Clark." Sprinkled throughout the article's detailed account of the great expedition are maps delineating the route of the explorers against the background of modern settlements, place names, and the rail route. Tourists following the railroad all the way to the Pacific would be privileged to see the same sites as Lewis and Clark; after a fashion, they would experience that pioneering journey anew.[1] Reading the article would have been a fine diversion on the long train journey, but the maps show that while the destination and general route were the same, the railroad tracks only rarely "literally followed the trail" (fig. 9.1). Passengers had no opportunity to get up close and explore. The technology mediated the jump from the present to the past only in passing, bringing the sites at which events took place into view only as the fixed route of the main line permitted.

A century later, at the bicentennial of the expedition, a flurry of guidebooks allowed motorists to retrace the Lewis and Clark trail more closely and at

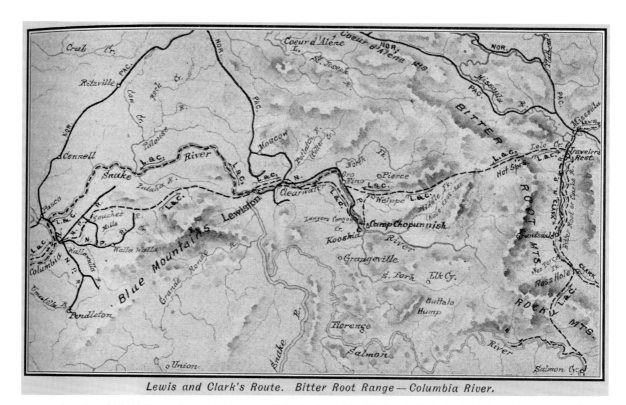

FIG. 9.1 "Louis and Clark's Route. Bitter Root Range—Columbia River," in Olin Dunbar Wheeler, "On the Trail of Lewis and Clark," *Wonderland 1900* (St. Paul, MN: Northern Pacific Railway, 1900). Image courtesy of the Newberry Library. The passage through the Bitter Root Range was one of the most difficult the Corps of Discovery faced. As this map shows, however, despite Wheeler's claims that the railroad closely followed the Lewis and Clark route, the mainline of the Northern Pacific actually traversed the range well to the north.

their leisure. One of the more successful was *Along the Trail with Lewis and Clark*, written by Barbara Fifer and Vicky Soderberg with excellent maps by Joseph Mussulman. The 215-page softcover guidebook followed the protagonists from well east of the Mississippi to the mouth of the Columbia and back again, with Mussulman's cartography helping modern motorists find their way along the route. He incorporated dated journal excerpts from Lewis, Clark, and other members of the Corps of Discovery, using arrows to pinpoint the location where the comments were made or to which they referred. Map 17A (fig. 9.2) focuses on the "Rocky Mountain Challenge," the party's search for a suitable westward path from the crest of the continental divide along the modern Montana-Idaho border.[2] As a comparison of figures 9.1 and 9.2 suggests, automobiles and highways had made it possible for historically minded travelers of the late twentieth century to make a more thoroughgoing—and more spatially explicit—imaginary leap into the past than had been possible a century earlier.

The historical tourist's imagination of the past is also complicated by overlapping, local, national, and even transnational collective imaginations of places and events. For Americans, the Lewis and Clark Expedition is an event deeply engrained in the collective national memory of transcontinental expansion and conquest, sustained by generations of school history lessons and ambient popular cul-

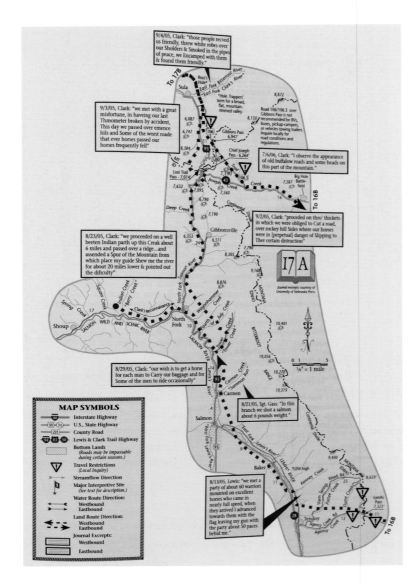

FIG. 9.2 "Rocky Mountains Challenge," in Barbara Fifer and Vicky Soderberg, *Along the Trail with Lewis and Clark* (Helena: Montana Magazine and Farcountry Press, 2001), 127. Courtesy of the Farcountry Press, Helena, MT; from *Along the Trail with Lewis and Clark*. Image courtesy of the Newberry Library.

ture. This, of course, is why the Lewis and Clark centennial and bicentennial were the occasion for map and guidebook publications and why historically minded American tourists might be inclined to invest the time to read and then follow these guides. In so doing, they are in some sense reaffirming their membership—to use Benedict Anderson's term—in the imagined community of the United States.³ So it is with much historical tourism of the nineteenth and twentieth centuries. The Grand Tour that reinforced pan-European identity through the study in place of European culture and history had not, by any means, faded away, but over the course of the nineteenth and early twentieth centuries it competed with, and increasingly yielded to, the visitation to sites and landscapes of later events. This was especially true in the United States, where the urgency of forming a distinctive national identity conflicted with the enduring fascination with ancestral European culture and history. Thus nationalistic American

Time, Travel, and Mapping the Landscapes of War [195]

tourism often focused on what were seen as distinctly American landscapes, Niagara Falls, the Hudson and Mississippi rivers, and what became the primary content of the national parks, notably in the West. But by the turn of the twentieth century, historic sites and foundational events also contributed to the content of American national tourism, none more so than the fields of battle.[4]

Just as the conduct of war has, throughout modern history, necessitated the production of a massive cartographic archive, so have the geographical complexities of battlefields and campaigns defied the comprehension of battlefield tourists, in ways that would seem to call for purpose-made maps. The creation of maps for military tourists has nevertheless had an uneven history, subject to the availability of good sources and especially the influence of technology. Battlefield tourism has always posed particular challenges to cartographers, since in addition to conveying standard features such as topography, they have the added burden of conveying action (often including a series of complicated advances and retreats across a given landscape). The successful map must direct the tourist's imagination toward the action of battle to the fullest extent possible while grappling with two problems: that the action has passed—it is not present—and that the landscape by which one might recall that past has changed over time. As we shall see, battlefield cartographers have tackled these challenges in a variety of innovative ways, incorporating text and image into increasingly complex multimedia assemblages.

The present essay offers a preliminary survey of the representational tactics developed by guidebook authors and mapmakers over the past two centuries for battlefield tourism,[5] a particularly robust form of what we might call "time traveling": travel directed at developing insights about past events by visiting the terrain where they unfolded. Though the focus here is on North American battlefields and American maps, comparisons with similar developments in Europe are inevitable. The story properly begins, in fact, in the aftermath of the Battle of Waterloo in 1815, where modern battlefield tourism and mapping for tourists had its first glimmer. Thereafter, I focus on how mapping for tourists responded to three conflicts: the War of 1812, the American Civil War, and World War I. All three were important inflection points in the evolution of historical tourism. The rise of travel publishing in the United States roughly coincided with the end of the War of 1812, and may be said to have been enabled by its outcome; the vast casualties of industrial warfare—and the parallel rise of the railroad—made the Civil War a watershed in the rise of battlefield tourism as a global (not just American) phenomenon; and the Great War, coinciding with the dawn of automobility, transformed war-related tourism again, making it at once a mass activity and an essentially private one. The cartographic archive reflects each of those changes.

European Origins: Waterloo

Mass tourism to battlefields and memorials may be a modern phenomenon, but the practice of making pilgrimages to militarily significant sites is one with

deep roots in European culture. The seventeenth- and eighteenth-century Grand Tour was designed, among other things, to give English aristocrats a place-based familiarity with the foundations of continental culture. The itinerary entailed many visits to landscapes of war, whether medieval castles, the sites of ancient battles, or monuments dedicated to the glory of Roman conquerors.⁶ The increased engagement of national populations in warfare during the Napoleonic Wars, however, marked a turning point in the way wars would be remembered—and by whom. The Battle of Waterloo (1815) in particular redirected battlefield tourism from the deep to the recent past, while bringing many more travelers onto the scene.

Such was the British public's interest in the site of Napoleon's catastrophic June 1815 defeat at Waterloo in present-day Belgium that the battlefield almost instantly attracted tourists. Within weeks of the battle, a local cottage industry had already emerged to serve them. Of his journey to the field on July 31, 1815, James Simpson wrote that his party passed fresh and "imperfect" graves: "Often bayonet scabbards stuck out; and caps, shoes, and pieces of cloth, scarcely in the gloom distinguishable from the mud in which they lay." Soon, they "were surrounded by the people offering for sale, with great importunity, relicks of the field; particularly the eagles which the French soldiers wore as cap plates."⁷ A year later, Seth William Stevenson reported that the graves along and around the approach to the battlefield and local memories of the battle were still fresh, and the cottage industry serving tourists booming. Taking breakfast at an inn and chatting with its keepers, he found that

> these people acknowledged the Battle of Waterloo had been of "*some service*" to them.... [and] we needed no Ghost to rise to inform us *they were doing very well*—for not less than half a dozen carriages were at the door, and parties continually arriving, who took refreshment and proceeded on.⁸

The physical environment around Simpson nevertheless enforced a somewhat uncomfortable sense of the contrast between the past and the present.

> A peculiar and forcible impression took place in our minds, excited by the mere reflection which occurred at the relation of the facts connected with the awful moment, when the habitation in which we were, stood within the region of death and on the verge of destruction, as contrasted with the tranquil security in which we now put our interrogatories, eating the while our rasher of *broiled ham*, and washing it down with the wholesome and pleasant *bierre brune de Louvain*, in the clean little parlour....
>
> With a very youthful, but ... sufficiently intelligent guide in our company, and an excellent map in out hand, we set out on our expedition over the field of battle.⁹

We cannot know what map they had in hand. Perhaps it was a copy of one published in 1816 in Brussels by William Benjamin Craan, an engineer in

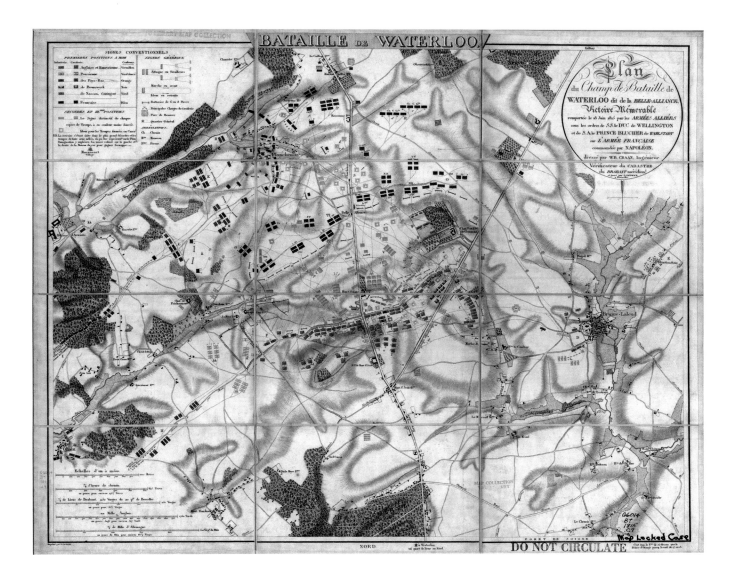

the service of William of Orange. Craan's large-scale annotated map, based on his own survey of the field and interviews with combatants, was widely reproduced for decades (fig. 9.3).[10] Neither Simpson's nor Stevenson's accounts included maps, relying instead on the text to describe the topography of the battlefield and their imagination of the events and people that previously occupied it. Only in the 1840s would the guidebook publisher Edward Murray combine descriptions of the battle with small woodcut maps of the field.

Mapping the Second American Revolution

The decades following the War of 1812 coincided with the birth of modern leisure-tourism in both Europe and the United States.[11] The industrialization of travel, via steamships and locomotives, in turn spawned an industry to supply commercial guides for travelers. The first modern guidebook series, published by the British firm of Edward Murray and the German firm of Karl Baedeker, appeared in the 1830s. Thomas Cook is said to have founded the modern commercial group-tour in 1841.[12] These developments

FIG. 9.3 William Benjamin Craan, *Plan du champ de Bataille de Waterloo* (Brussels: W. B. Craan, 1816). Wikipedia Commons.

coincided with rapid improvements in the reliability and speed of travel, thanks to passenger railroads and canal boats.

In the United States, commercial maps and guidebooks catered not only to tourists but also to migrants. Though indecisive militarily, the War of 1812 had decisively settled one question: the British and their indigenous allies would not be able to slow or reverse American settlement in the trans-Appalachian West.[13] With their northern frontier effectively pacified, free white Americans could now move between the Ohio River and the Great Lakes with few limitations beyond those of their own will and resources. The first national guidebooks and maps for travelers were published in the immediate aftermath of the war; by the late 1820s, multiple publishers had entered the market.[14]

Guidebook authors made a point of highlighting points of historical interest, even when addressing migrants. Zadok Cramer's popular 1821 guide to the Ohio and Mississippi rivers, for example, directed the eyes of his readers to a point on the Allegheny River north of Pittsburgh, where British general "Braddock's army was cut to pieces in a short time; the Indians and French being completely covered by woods, grass, &c . . ." Cramer reported that the contemporary owner of the site "frequently ploughs up the bones of those who fell in this dreadful carnage."[15]

Publications oriented specifically to tourists emphasized historical themes more heavily. In the early years, the focus of American travel literature was the scenic landscapes of New England, New York, and (thanks to the peace with Great Britain) the St. Lawrence Valley. Guidebooks promoting this "northern tour"—pitched to visiting Europeans, wealthy Easterners, and the southern planter aristocracy—emphasized the scenic wonders of the Hudson River, the spas around Saratoga, Lake Champlain, the White Mountains, Quebec, and Montreal, and above all, Niagara Falls.[16] It is among these accounts that we find the first consistent references to landscapes of war. The campaigns along Lake Erie, Lake Ontario, and the upper St. Lawrence River received particular attention, nowhere more so than at Niagara. Three major battles—Queenston Heights (October 13, 1812), Chippawa (July 5, 1814), and Lundy's Lane (July 25, 1814)—were contested on the Canadian side of the river within a few miles of the Falls and were easily reached by excursions. A monument honoring British major general Sir Isaac Brock, who died on Queenston Heights, was constructed in 1824 and became a touchstone of many guidebook accounts.[17]

In his *Tour from the City of New-York, to Detroit, in the Michigan Territory* (1819), William Darby remarked extensively on the still-recent conflict. He also mused on the poignant disappointments of his tour: "There is . . . no scene which the traveller visits, that so little answers his expectations as the field of battle. . . . The eye finds nothing beyond the common objects in nature to render conspicuous the scene of the greatest battles."[18] Darby's detailed map, "The Straits of Niagara," clearly marked the location and dates of the battles of Queenstown, Chippewa, and Lundy's Lane, and the locations of Ft. George and Ft. Erie, at either end of Niagara River. A second map, of the "Environs of Detroit," did not mark the land

battles in the area (which were mostly American defeats), but it did mark the location in Lake Erie of Commodore Matthew Perry's naval victory on September 10, 1813. Darby expressed his regret that he could only glimpse the latter from a distance, due to a stiff west wind.[19]

For American guidebook publishers during this period—at least those who treated the Niagara region—battlefields were merely places to be noted (literally) in passing, diversions worth a pleasant excursion for those who were historically inclined, but not the main attraction. Most maps appearing in guides to Niagara in the 1830s simply marked the sites of the battles. In the 1840s and 1850s, by contrast, the development of lithography and advances in wood engraving made it possible for publishers to add more detailed graphics.[20] In the United States and Europe, readers could now encounter battles with unprecedented immediacy through illustrated magazines that published views and maps of battles that had been drawn in the field.[21] If the *Illustrated London News* and France's *L'Illustration* helped readers visualize distant imperial conquests or relive the ferocious character of the Crimean War (1853–1856), guidebook series such as Murray's, Baedeker's, and the French *Guides Joannes* likewise expanded their repertoire of views and maps.[22]

In the United States, the historian Benson J. Lossing embraced the extensive use of topographical imagery to help readers make the leap from the present landscape to historical action. Lossing began his career as a literary publisher in New York. An accomplished sketch artist, he taught himself the art of wood engraving, a popular technique in the mid-nineteenth century for the production of inexpensive images that could easily be combined with printed text. Over a career spanning nearly five decades, Lossing published more than forty books, including several illustrated biographies and histories—all notable, despite his lack of formal training, for diligent research and extensive personal interviews.

Among Lossing's greatest achievements were two massive and richly illustrated studies, *Pictorial Field-Book of the Revolution* (1851–1852) and *Pictorial Field-Book of the War of 1812* (1868).[23] The two works are nearly identical in style, and distinguished from his other historical works by the term "field-book" in their titles. Illustrated histories they certainly are, but Lossing emphasized that most of the views and maps and much of the narrative were the result of his own observations, including interviews with combatants and eyewitnesses. "The author has traveled more than ten thousand miles in this country and in the Canadas, with note-book and pencil in hand, visiting places of historic interest connected with the War of 1812, from the Great Lakes to the Gulf of Mexico, gathering up, recording, and delineating every thing of special value, not found in books, illustrative of the subject, and making himself familiar with the topography and incidents of the battlefields of that war."[24] The term "field-book" traditionally referred to a notebook in which land surveyors marked and kept their measurements, descriptions, and sketches as they went about their work. By the mid-nineteenth century, its meaning had broadened to include books of sketches kept by artists and naturalists (especial-

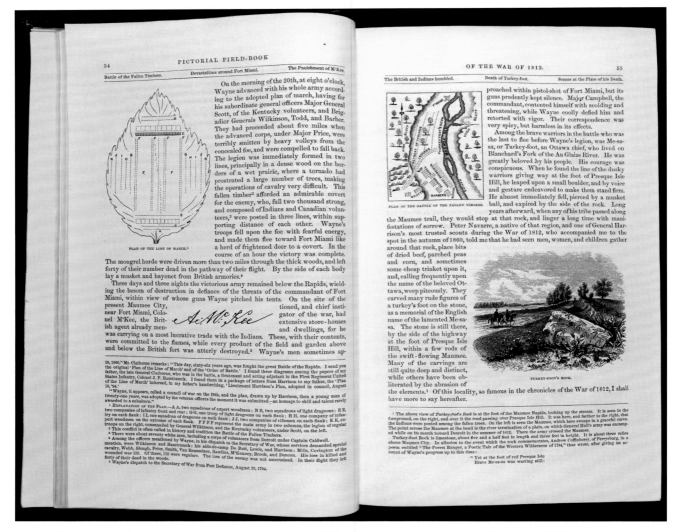

FIG. 9.4 Benson Lossing's sketches and descriptions of the Battle of Fallen Timbers, in Benson J. Lossing, *The Pictorial Field-Book of the War of 1812* (New York: Harper & Bros., 1868), 54–55. Image courtesy of the Newberry Library.

ly botanists and zoologists) to record observations made outdoors. Lossing's usage here referring to notes and sketches made in the theaters of war was therefore new, though not unique.[25] He makes a point of inserting himself as a traveler and observer, interrupting the historical narrative with brief accounts of his own travels and interviews.[26] Here is his account from *The War of 1812* of his visit to the site of Fallen Timbers (1794), the climactic battle in the first post-Revolution Indian war in the Northwest Territory:

> Among the brave warriors in the battle who was the last to flee before Wayne's legion, was Me-sa-sa, Turkey-foot, an Ottawa chief, who lived on Blanchard's Fork of the Au Glaize River. He was greatly beloved by his people. His courage was conspicuous. When he found the line of dusky

people giving way at the foot of Presque Isle Hill, he leaped upon a small boulder, and by voice and gesture endeavored to make them stand firm. He almost immediately fell, pierced by a musket ball, and expired by the side of the rock. Long years afterward, when any of his tribe passed along the Maumee trail, they would stop at that rock, and linger a long time with manifestations of sorrow. Peter Navarre, a native of that region, and one of General Harrison's most trusted scouts during the War of 1812, who accompanied me to the spot in the autumn of 1860, told me that he had seen men, women, and children gather around that rock, place bits of dried beef, parched peas and corn, sometimes some cheap trinket upon it and, calling frequently upon the name of the beloved Ottawa, weep piteously. They carved many rude figures of a turkey's foot on the stone, as a memorial of the English name of the lamented Me-sa-sa. The stone is still there, by the side of the highway at the foot of Presque Isle Hill within a few rods of the swift-flowing Maumee.[27]

A map of the engagement and an engraving of Me-sa-sa's rock accompany this passage (fig. 9.4). In Lossing's description we can also hear the faint cross-cultural echo of the memorial landscape practices of Native Americans, who Lossing claimed returned to the site "long years afterward" to commemorate it.[28]

Lossing's field-books demonstrate the potential of carefully coordinated text, image, and map to support the time-traveling battlefield tourist. But they are not guidebooks; they offer no practical guidance to routes or accommodations. What they offer are invitations to the imagination, designed to help armchair travelers envision and recreate the scenes of battle. It would take another war of rebellion before publishers began creating guides specifically designed to direct the movement of travelers to and through the fields of battle.

Mapping and Moving through Landscapes of the Civil War

The American Civil War and its aftermath provided the ideal circumstances for the emergence of battlefield tourism as a mass phenomenon. Since the Napoleonic Wars, with their mass conscription of national armies, the numbers of the dead (and consequently of aggrieved families) had dramatically increased. If European states by mid-century had begun to organize the transport and burial of war-dead in select locations, the sheer number of fatalities at Shiloh, Fredericksburg, Antietam, and Gettysburg necessitated the rapid creation of cemeteries on site. It was to dedicate one such soldiers' cemetery that President Lincoln composed his famous Gettysburg Address. The presence of this official gravesite stimulated the battlefield's attraction as a place of pilgrimage. As at Waterloo, even before the war was over the local population had begun to serve as makeshift hoteliers and informal guides.[29] Professional tour-leaders and printed guidebooks followed soon after.

To an extent far greater than in any previous North American conflict, the American Civil War engaged a large portion of the US population, wheth-

er directly or indirectly. In all, more than 2.6 million men enlisted in the Union Army and about one million served in the Confederate forces. Casualties for the combined armies exceeded 1.1 million, of which more than 200,000 were killed in battle. Fought on land and water, the war spurred the development of automatic weapons, artillery of increasing accuracy and explosive power, submarines, and armored warships and gunboats. It was perhaps the first major global war in which railroads were used extensively to move troops to the front. After the war, these same railroads made it possible for large numbers of combatants to revisit the battlefields. Veterans, their families, mourners, and the curious toured the battlefields in unprecedented numbers, creating markets for professional guides and prompting the creation of the first guidebooks and maps explicitly designed for battlefield tourists on an equally unprecedented scale.[30]

No previous war was documented so extensively by visualization, including the new medium of photography. Photos and field sketches from the front (including those of Benson Lossing) were published regularly in illustrated periodicals, and both *Leslie's* and *Harper's* were quick to publish illustrated histories repackaging these images.[31] The demand for imagery extended to maps, which appeared in daily newspapers in major northern cities. Most were either gleaned by war correspondents from military maps and sketches or taken on the spot by civilian eyewitnesses if not penned by the journalists themselves.[32] Separate sheet maps documenting the progress of the war were also rushed to print.

The visualization of the war continued with renewed intensity after the conflict. The sheer size of the armies engaged invited those who sought to capture battlefield action to engage with the landscape on a panoramic scale. Public displays of outsized paintings of historical interest had been a popular form of entertainment since the late eighteenth century, in both Europe and America. These included both conventional panoramas (which typically presumed a single-point perspective) and cycloramas (circular presentations with multiple vanishing points).[33] Probably the most famous of the latter in the nineteenth century was French artist Paul Phillippoteaux's cyclorama, nearly 300 feet in diameter, depicting Pickett's Charge, the decisive moment from the third day of the Battle of Gettysburg (fig. 9.5). First displayed in Chicago in 1883, the original was featured at Chicago's World Columbian Exposition ten years later and again at the St. Louis World's Fair of 1903. A second copy traveled from Boston to several other cities before being permanently housed at the Gettysburg Visitor Center. The genius of the cyclorama was to put the viewer's body in the middle of the action—in a manner that anticipated 3-D films and virtual reality. Wrote the *Chicago Tribune*:

> The battle ground, with its dead and wounded soldiers, the smoke of cannon, the bursting of shells, the blood stained ground are all drawn with a realism that is almost painful. The spectator can almost imagine that he hears the rattle of musketry and the brave regiments as they charge upon each other to sink amid the smoke and carnage . . . it

FIG. 9.5 Index map of Paul Phillipoteaux's cyclorama of Pickett's Charge, in *Gettysburg* (n.p.: Union Square Panorama Company, 1886?). Image courtesy of the Newberry Library.

taxes the ingenuity of the looker on to tell where the real ends and where the work of the brush begins.[34]

This panoramic simulacrum was instrumental in the most popular early guidebook to the battlefield, John B. Bachelder's *Gettysburg: What to See, and How to See It* (1873).[35] Bachelder was an artist with a military background who attached himself to the Union Army of the Potomac in 1862, hoping that he might record in word and image the decisive battle of the war.[36] When it became evident that Gettysburg might be that decisive battle, he surveyed the

FIG. 9.6 John Bachelder, "Gettysburg Battlefield. Battle Fought at Gettysburg, Pa., July 1st, 2d & 3d, 1863 by the Federal and Confederate armies," in Bachelder, *Gettysburg: What to See, and How to See It* (Boston: Bachelder, 1873). Image courtesy of the Newberry Library.

field over a period of three months, sketching and recording observations that would allow him to produce an "isometric map" that collapsed the events of the three-day struggle into a single image. Later in 1863 he published his panorama with signed endorsements from General George Meade and other Union commanders.[37] As with Lossing's field-books, Bachelder's panorama relied heavily on eyewitness testimony, which he continued to gather. In 1864 he published a *Key to Bachelder's Isometrical Drawing of the Gettysburg Battle-field, with a Brief Description of the Battle*; within a decade, the *Key* had been expanded into a full-fledged guidebook.[38]

Bachelder's guidebook featured a monochrome version of his "isometrical drawing" overlaid with a grid, referenced to an alphanumeric index that allowed readers to locate places and actions described in the text. He advised his readers to begin their tour of the battlefield by visiting the theological seminary that gave its name to Seminary Ridge (D-12 on the panoramic map; fig. 9.6). As he did throughout the book, Bachelder described at length the disposition of troops on the eve of battle as it might have been seen from the highest point in the terrain (here, the cupola of the seminary). To further aid the tourist in imagining the events that unfolded at their feet, Bachelder here and at subsequent viewpoints inserted a drawing of a compass (fig. 9.7), orienting the viewer to key landscape features described in the text.

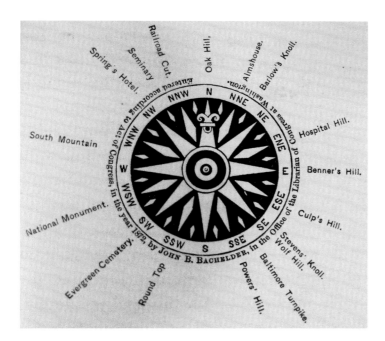

FIG. 9.7 John Bachelder's "Compass-Face Guide" for the Gettysburg Theological Seminary, in *Gettysburg: What to See, and How to See It* (Boston: Bachelder, 1873). Image courtesy of the Newberry Library.

By carefully consulting this, as the reader finds it presented from time to time, he will be at no loss in pursuing his studies. For example, if standing upon East Cemetery Hill, you turn that point of the Compass-Face Guide marked Theological Seminary towards that object, the National Monument, Culp's Hill, and all other places named on the diagram will point directly towards the real objects; and the result will be the same at any place where it is presented, if you turn the name of any one known locality towards the object itself.[39]

Bachelder's maps and guidebooks were the first of many devoted to Gettysburg published in the ensuing decades. Bachelder asserted that the best way to tour the battlefield was on horseback, as both he and military commanders had surveyed it in 1863. Recognizing that this would be impractical for some travelers, he closed the guidebook with "driving tours" that could be taken in carriages. Gettysburg's location at a crossroads—the root of its military significance—meant that it was easily reached by these conveyances. Rail travelers could reach it only from a single prewar railroad branch leading from Hanover Junction, thirty miles to the east. In 1884, however, a new railroad opened from Gettysburg to Harrisburg with a branch line to Round Top Park, an attraction and base for touring excursions operated by the railroad.[40] The lines leading to Gettysburg were consolidated as part of the Western Maryland Railroad in 1886, which published an illustrated guidebook to the town and battlefield in 1890.[41] An electric trolley that circled the battlefield was even opened in 1892. This popular attraction ran for twenty-five years, until preservation concerns and competition from automobiles finally shut it down.[42]

The railroad and the automobile were great enablers, but the battlefield's popularity—as well as its development as a tourist site—rested initially on its emotional appeal to war veterans and their families. Here, as elsewhere in the battlescapes of the Civil War, the survivors of the battle sojourned over the next half century, informally or in reunions organized by their military units, memorial associations, and veterans' organizations such as the Grand Army of the Republic. Organized reunions and memorial dedications were major impetuses for the publication of maps and guidebooks.

In 1883, William H. Whitney, a brevet major from the 38th Massachusetts Volunteers, published *Union and Confederate Campaigns in the lower Shenandoah Valley Illustrated: Twenty years after*. This atlas of thirty-three maps, beautifully hand drawn, was reproduced in blueprint for "the first reunion of Sheridan's veterans on the fields and camps of the valley."[43] The compilation, based on maps from pre-

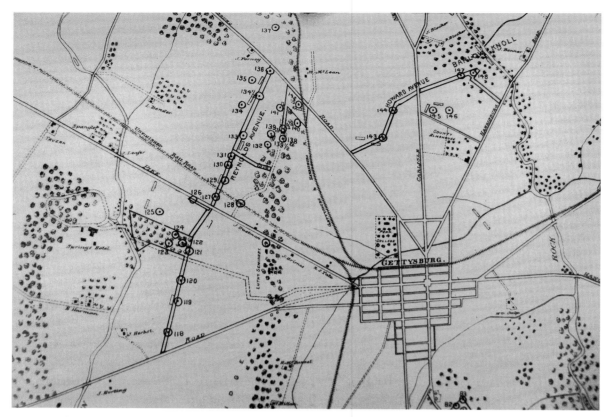

FIG. 9.8 Detail, Charles R. Graham, "Map of Gettysburg, Pa. and Vicinity, Showing the Positions of the Monuments to Be Dedicated during 1888," in John Tregaskis, *The Battlefield of Gettysburg: The Men Who Fought Here and the Monuments Dedicated* (New York: Tregaskis & Co., 1888). Image courtesy of the Newberry Library.

viously published histories, government sources, and his own reconnaissance, was motivated by the "special difficulty of retaining in the mind the events of many campaigns of the four years in that region." Most of the maps are large in scale, attending carefully to topographic details and troop movements; most include brief narrative descriptions of the actions. Though there are no direct instructions for the tourist, across the top of each map Whitney provided a blank box in which users might add their own "private records" of engagements in which they participated. Several maps also indicate where topographical changes might affect memories of the engagement. For example, on map 33, "Battlefield of Winchester, Va." (September 19, 1864), Whitney notes that the

"old road up this ravine [has been] ploughed over" and "new slat fence [appears where a] Virginia rail fence was" and "new barb wire fence."[44]

Monuments to individual units and commanders started popping up in the immediate aftermath of the war, accelerating in the 1880s and 1890s. Initially these were bottom-up efforts, many undertaken by individual regiments. Over time, state commissions stepped in to organize and fund these memorials, working alongside local battlefield commissions that regulated their placement. For the most part, monuments were erected at key locations where units were engaged. At the site of the largest engagements, such as Gettysburg, Antietam, and Chickamauga, these lined up to form "avenues," paths well suited for the

Time, Travel, and Mapping the Landscapes of War   [207]

movement of tourists retracing the course of battle. Battlefield commissions regulated the placement of monuments to ensure historical accuracy.[45]

This phenomenon is illustrated by a map prepared by Charles R. Graham for the Gettysburg Monument Commission of the State of New York "showing the positions of monuments to be dedicated during 1888 by all states on the occasion of the 25th anniversary of the battle in 1888" (fig. 9.8). Linear clusters of monuments along major points of conflict are labeled Sickles, Reynolds, or Hancock Avenue, after the Union generals who commanded these sectors of the line. (These names are preserved in the names of modern roadways that traverse the battlefield.) For veterans and other visitors, coordinating the monuments into avenues helped direct both movements and sight lines.

The 1890s brought an increased federal interest in the preservation of the battlefields. Battlefield commissions, which had been actively acquiring battlefield sites in order to control their preservation, now turned these lands over to the federal government for the establishment of national military parks. The combined sites of the Battles of Chattanooga and Chickamauga formed the first such park in 1890. The military park commissions created new maps to guide further development, but these were detailed documents designed for internal use.

Railroads, by contrast, had a financial interest in publishing maps and guidebooks to the new war-themed parks. The Chesapeake & Ohio—which was especially well positioned to transport visitors to battlefields in the mid-Atlantic region—put out a *Map showing the location of battlefields of Virginia compiled from official records and maps* in 1891, marking the sites of more than a hundred engagements (fig. 9.9). Each was assigned a number on the map, indexed on the back to the name and date of the battle. Acknowledging the role that the railroads themselves had played in war logistics, the map carefully distinguished between lines that were in service during the war, and lines that were constructed afterward. Attacks on the railroads themselves were also indicated ("Raid on Danville R.R."; "Raid on South Side R.R."). The major organization for Union veterans, the Grand Army of the Republic, issued a version of this map in 1906 as publicity for its annual encampment in Cincinnati. The accompanying commentary artfully unspooled the event horizon of 1865 at the pace of train travel:

> To [the veteran] journeying to the great encampment of his fellows, or to those who wish to visit the memorable fields of American fighting, the trip from Washington to Cincinnati by the Chesapeake & Ohio, is one of intensest interest . . . .
>
> The road crosses the Potomac on the Long Bridge and in sight of the Aqueduct Bridge—the two lines over which the living stream of men flowed with the advance of the Union army. . . . On reaching the Virginia shore, the ruins of the immense earthworks are about the end of the bridge and everywhere on the surrounding hills. The eye of the veteran will be quick to catch the outlines of parapet and embrasure against the sky. At Alexandria, the Marshall House, where Ellsworth

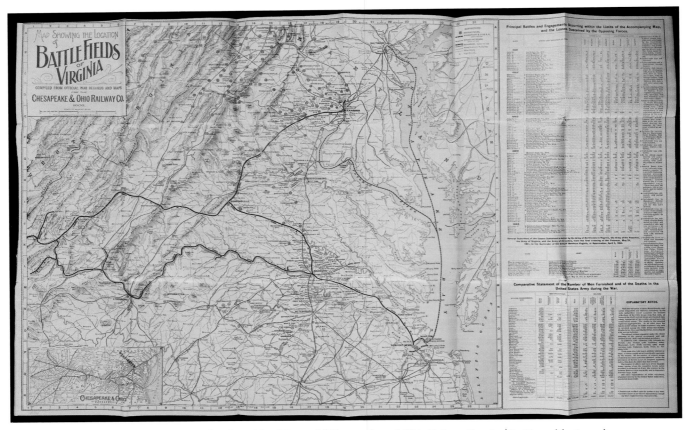

FIG. 9.9 "Map Showing the Location of Battle Fields of Virginia" (Chesapeake and Ohio Railway Co., 1891) in *Manual for Annual Encampments Issued by the Passenger Department, Chesapeake & Ohio Ry.* (Chicago?, 1906). Image courtesy of the Newberry Library.

was killed, is near the track. The tower of Fairfax Seminary, where the first skirmishing began in the advance to Bull Run, is in plain view. In half an hour the train is skirting the banks of that stream. In a few moments more a heavy confederate work is seen on the left; the train, just after passing, stops at Manassas, and the veteran is once again on the ground where the war for the Union began.[46]

This passage would be one of the last penned for living survivors; as veterans passed away, so too did guidebooks designed for those who carried personal memories of the war. While subsequent generations would find themselves at a greater temporal remove from the events of 1865, entirely new opportunities for exploring the terrain of war would be opened by the automobile.

The Great War and Battlefield Tourism in the Automobile Age

The possibilities of automobile-based battlefield tourism were first developed in the aftermath of World War I. The French tire manufacturer Michelin, which had been publishing touring maps and guidebooks for about a decade before the war, soon began to publish guidebooks suited to motor travel for Allied audiences at war's end.[47] As with the American Civil War, postwar battlefield tourism responded in part to the vast influx of grieving families and

Time, Travel, and Mapping the Landscapes of War

FIG. 9.10 Touring map of operations in the Aisne-Marne region, in *A Guide to the American Battle Fields in Europe* (Washington, DC: Government Printing Office for the American Battle Monuments Commission, 1927). Image courtesy of the Newberry Library.

comrades-in-arms, who flocked to the Western Front in search of loved ones' graves or simply to walk the ground where they had fought and died. Michelin's innovative outlines of automobile tours were decidedly for travelers who owned or could afford to hire a car or motor-enabled tour. One historian has argued that they were better suited to former officers (or noncombatants of means) than they were to common soldiers and their families, whose major perspective on the battle was from the trenches. But the sale of more than 1.4 million copies of these guides by 1922 would suggest otherwise.[48] Nevertheless, the guidebooks showed for the first time how automobile travel, mapping, photography, and narrative could be effectively combined to visualize the scenes of battle over entire battlefield sectors. For the battlefield tour described in the 1919 English edition of the Michelin guide to the First Battle of the Marne (1914), precise driving instructions ("Turn to the right at the foot of the slope into Étrépilly and on leaving the village take the road on the left") accompanied a detailed map of the region, dense with text and iconography related to the war. To these were added battle narrative and photographs ("In the counter-attacks, the Germans, as they left their trenches, also suffered serious losses, as one can judge from the photography above"), all keyed to the map.[49] In another example, showing a portion of the Western Front along the Ypres (Belgium) salient, small monochrome maps of portions of a prescribed route are laid into navigational, narrative, and descriptive text, along with photographs and panoramas designed to help readers survey the battlefield from many angles and points of view.

Michelin published twenty-nine separate titles for the Western Front in the first decade after the war, some appearing in English translation and designed to appeal to European and British Commonwealth travelers. American-published guidebooks compensated by focusing on the sectors of the Western Front where Americans served. *America in Battle* (1927), by Harry S. Howland and James A. Moss, described eight detailed itineraries of the American sectors of the Western Front. The five accompanying maps were entirely historical in character. Rather than incorporating navigational aids into their cartography, the authors advised visitors to study in advance and to bring their powers of imagination to bear: "With mind and imagination properly attuned to the drama of war you may visualize all the moving elements of conflict."[50]

By contrast, *A Guide to the American Battle Fields in Europe* (1927)—published by the American Battle Monuments Commission—managed to overlay maps of the battle sectors with maps outlining tours. The chapter entitled "Operations in the Aisne-Marne Area" opened with an eleven-page history of American operations in that sector, illustrated with numerous photographs and two maps. The tour that followed occupied thirty-three pages, richly illustrated with photos and panoramas of the front as the tourist would find it. The detailed itinerary included instructions on precisely where to turn, where to pause, and where to look.[51] At the end of the chapter, a folded four-color map prepared by Army engineers offered an overview of the entire tour, along with some alternate routes, overlaid by the lines marking locations

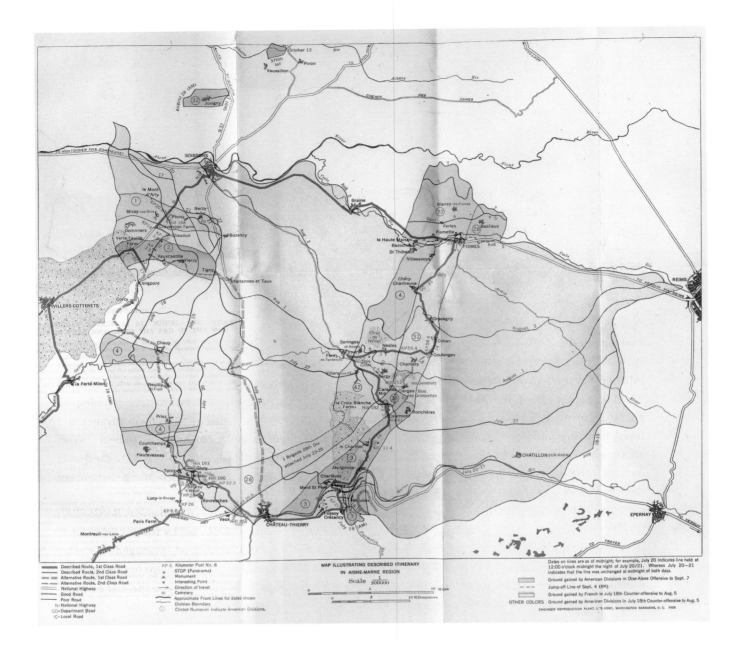

of American divisions (fig. 9.10). The preface relates that the elaborately researched and produced book was "expedited in order to have it available for the large number of ex-service men who intend to go to Europe in the fall of 1927," the tenth anniversary of the American entry into the war.

The 1920s coincided with the spectacular growth of automobile ownership in the United States, and with it the general social expansion of automobile tourism among the middle and working classes. By the middle of the twentieth century, the vast American highway infrastructure and associated service industries provided a broad spectrum of Americans in all regions of the country with the means to travel

FIG. 9.11 *1862 Antietam Campaign: Lee Invades Maryland* (Annapolis?: Maryland Civil War Trails, 2015). Courtesy of the Civil War Trails, Williamsburg, VA. Image courtesy of the Newberry Library.

long distances, at their own pace, following itineraries of their own design. The common American road map enabled this self-guidance. But local and regional competition for tourist dollars also encouraged the production of more specialized cartography to direct drivers in specific ways. Many of these publications resorted to patriotic rhetoric, casting motor travel as an activity capable of bringing Americans in touch with their vast country and its history. The experience of the Great War had encouraged American travelers to look inward, to "See America First," and much of this involved a renewed interest in sites associated with conflicts such as the American Revolution and Civil War that fit well into the paradigmatic national narrative of rising freedom and glory.[52]

The majority of auto-oriented maps of war-related sites simply called attention to battlefields, showed their locations, and offered directions to them. But the tourist mapping of Western Front landscapes of war had shown that automobiles enabled travelers to closely follow the course of past events in the present, alternating between the ground trod by common soldiers and the panoramic views afforded their commanders. What was new was both the motorcar's speed and the increasingly built-up landscape through which it traveled—a landscape that was being steadily transformed by car culture. Bachelder's horse could be dispensed with, but at a cost. To an extent that Lossing and Bachelder may have envied, guidebook authors and mapmakers addressed audiences of battlefield tourists who could really cover ground with few impediments—with their pathways paved for them, in fact. Buohl's *Illustrated Map and Guide to Tour the Gettysburg Battlefield* outlined a driving tour that follows the action semi-chronologically, heading west from Gettysburg along the Chambersburg Pike toward the site of the first skirmishes. From there the recommended itinerary follows the lines of battle west and north of the town, where Confederate forces attacked and forced Union forces (and arriving reinforcements) to take up positions in the highlands to the south. The route then climbs those heights, tracing the entire Union line from Culp's Hill to Cemetery Ridge to Little Round Top, before descending to the Wheatfield and Peach Orchard—sites of intense fighting on the second day of the battle—and passing by the site from which Pickett's Charge was launched. Similar touring strategies tracing the lines of battle were promoted by other Gettysburg maps and guidebooks, including those published by the National Park Service. This strategy could also be applied effectively to the mapping and touring of entire theaters of war.

The cartographic innovations of the early automobile age are still with us. Just a few years ago, the Civil War sesquicentennial of 2015, coming fast on the heels of the bicentennial of the War of 1812, produced similar maps, albeit with greater refinement. For instance, the state of Maryland in collaboration with the Federal Highway Administration published a series of five colorful folded brochures devoted to the mid-Atlantic campaigns of the Civil War.[53] The brochure titled *1862 Antietam Campaign: Lee Invades Maryland* includes a large, clear map marking out a recommended itinerary that follows the course of the campaign (fig. 9.11). According to its designers,

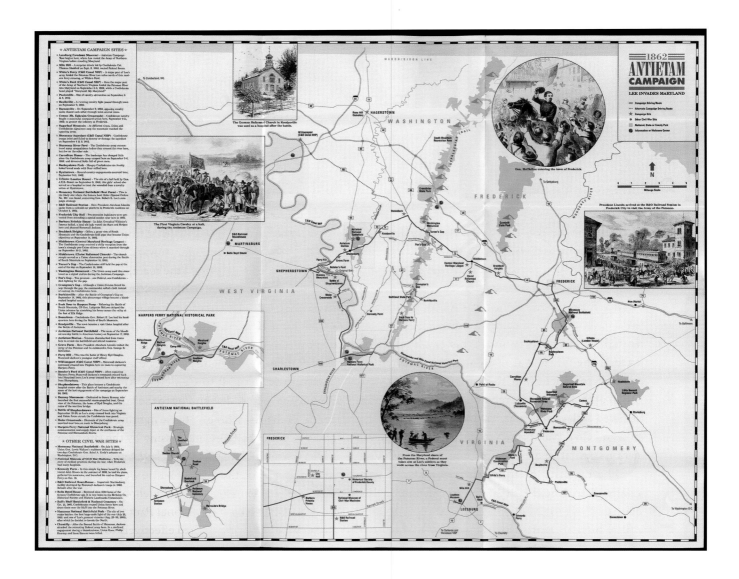

"This guide depicts a 90-mile historic and scenic driving tour that follows the route taken during Robert E. Lee's September 1862 Maryland Campaign. The tour reveals stories that have been hidden in the landscape for over 140 years." A blue line on the map traces the course of the campaign, connecting sites marked with red stars in chronological order. This well-executed traditional example of war-related time traveling, however, utilizes twenty-first-century innovations. For more information, readers are directed to the Civil War Trails official website (www. civilwartrails.org), where one can download "the free Maryland Civil War Trails mobile app to use as a companion to this guide while navigating the trail." Travelers are urged to "follow the bugle trailblazer signs to waysides that explain the day-to-day lives of soldiers and civilians as thousands of men and boys marched towards their undeniable destiny."

The bicentennial commemorations of the War of 1812 in Canada and the United States likewise generated a wave of histories, guidebooks, and maps, especially in Canada. Among these, the Canadian

Time, Travel, and Mapping the Landscapes of War [213]

1812: A Guide to the War and Its Legacy illustrates the potential for time traveling through the landscapes of war by making use of still newer technologies: aerial and satellite imagery.[54] Each of the seven chapters of the book's touring section features multiple satellite images or aerial photographs, annotated so as to allow travelers to follow the movements of past militia while referencing present-day landmarks. The narrative also provides the street addresses of key sites so motorists can find these places on their GPS. Typical pages accompanying a visit to the Battle of Lundy's Lane (outside Niagara Falls, Canada) combine an annotated satellite map of the Niagara vicinity with enlargements of specific sites, each keyed by letter to an associated textual passage and GPS location. The narrative for "Site D," for example, instructs tourists to travel from Lundy's Lane

> towards Fort Erie via the Niagara Pkwy. On your way you may wish to stop at Chippawa Battlefield Park and National Historic Site in Chippawa Village. (GPS—8709 *Niagara Pkwy. Niagara Falls, ON*). When you cross the bridge take a look at the river which in 1813 provided a significant natural barrier to an American advance to Fort George. Major-General Phineas Riall's decision to abandon this position, cross the river, and confront the Americans in an open field defies easy explanation.[55]

One might ask whether paper maps and guidebooks will be deployed to map historic time travel at all in the future. I suspect they will be, though the experience of imagined place and time as prescribed by Lossing and Bachelder will have changed. Whether time traveling will lose something from the virtual experience of place is hard to say.

Conclusion

Throughout this brief tour of mapping deployed to support "time travel" through the landscapes of war, we have resorted, like tourists, to the reading of texts to make plainer what guidebook and map authors had expected travelers to see and experience, both while approaching and while moving through the fields of battle. It is mostly in texts, adjacent to or incorporated into maps, that authors make explicit their expectations of what visitors must know to understand what they see. The maps considered here required text to animate the imagination of would-be time travelers. For the historian, those texts are now invaluable in their own right, helping modern interpreters of these maps understand the perspectives of both their authors and their presumed contemporary readers. We have also seen how text, photographs, and sketches were mobilized to provide testimonials that the authors—if not eyewitnesses to the events themselves—had taken pains to support their observations and descriptions with firsthand testimony, either by interviewing eyewitnesses or by quoting at length from their written accounts. Eyewitnesses simultaneously provide authenticity to the tour guide's narrative and help the reader make a formidable imaginative leap backward in time. Battlefield maps, in other words, are just one part of a multimedia tool-

kit aimed at war tourists; cartography was essential, but it was never meant to operate alone.

Over the course of the nineteenth and twentieth centuries, maps took on increasing importance in supporting the battlefield tourist's imagination of past time and place. Maps became cheaper to produce just as passenger trains, automobiles, and leisure travel increased demand for them. Maps were no longer mere adjuncts to essentially narrative accounts, but became critical tools for directing tourists' vision. With the era of the automobile, mapping took on the additional role of directing the movement of tourists' bodies around battlefields, recreating the maneuvers of military units and enabling rapid transitions from viewpoint to viewpoint. As tourists became more mobile, mapmakers could plan excursions that retraced the movement of armies and soldiers more thoroughly and explicitly. As a result of these transformations, battlefield time travel—once a leisurely activity practiced by a select (and highly literate) few—has become a mass phenomenon, deploying the latest technologies of both movement and image-making to enhance tourists' illusion of actually being there.

But "being there," as this essay has shown, is complicated in the case of battlefield tourism. This form of historical cartography is ultimately extremely selective in its temporality. The traveler to a former war zone reaches back to one particular moment in time while ignoring and even resenting all the living that may have transpired and transformed that landscape between the time of the battle and the time of the traveler. Battlefield tourism also brings into sharp relief the tensions inherent in commemorating death in the midst of life. The mundane and even profane realities of modern mass tourism—the crowds, noise, packaging, and commercialism—can conflict with the tourist's hope to make a dignified pilgrimage to ground consecrated by human blood. Finally, the conflict over the relative claims of past and present can bring the national battlefield commissions—intent on memorializing a version of the past—into conflict with local neighborhoods serving the needs of the present. In all these ways the battles of the past continue to echo in the present, and new soldiers are mustered to wage war on the fields of national memory.

Notes

1. Olin Dunbar Wheeler, "On the Trail of Lewis and Clark," in *Wonderland 1900* (St. Paul, MN: Northern Pacific Railway, 1900), 1–76.

2. Barbara Fifer and Vicky Soderberg, *Along the Trail with Lewis and Clark* (Helena: Montana Magazine and Farcountry Press, 2001), 125–26. A similar guidebook, but intended to encourage general touring around the route of the trail, is Kira Gale, *Lewis and Clark Road Trips: Exploring the Trail across America* (Omaha, NE: River Junction Press, 2006).

3. Benedict Anderson, *Imagined Communities: Reflections on the Origin and Spread of Nationalism*, rev. ed. (London and New York: Verso, 1991).

4. On the relationship between nationalism and tourism, with particular reference to the United States, see J. Valerie Fifer, *American Progress: The Growth of Transport, Tourist, and Information Industries in the Nineteenth-Century West Seen through the Life and Times of George A. Crofutt, Pioneer Publicist of the Transcontinental Age* (Chester, CT: Globe Pequot, 1988); John F. Sears, *Sacred Places: American Tourist Attractions in the*

Nineteenth Century (New York: Oxford University Press, 1989); Michael Kammen, *Mystic Chords of Memory* (New York: Alfred A. Knopf, 1991); Marguerite S. Shaffer, *See American First: Tourism and National Identity, 1880–1940* (Washington, DC: Smithsonian Institution Press, 2001); and James R. Akerman, "Twentieth-Century American Road Maps and the Making of a National Motorized Space," in Akerman, ed., *Cartographies of Travel and Navigation* (Chicago: University of Chicago Press, 2006), 151–206.

5. My understanding and usage of this term have been heavily influenced by David Wharton Lloyd, *Battlefield Tourism: Pilgrimage and the Commemoration of the Great War in Britain, Australia, and Canada* (Oxford: Berg, 1998).

6. Among the many works on the Grand Tour and its accounts are: Jeremy Black, *The British and the Grand Tour* (London: Croom Helm, 1985); Lynne Withey, *Grand Tours and Cook's Tours: A History of Leisure Travel, 1750–1915* (New York: W. Morrow, 1997); and James Buzard, "The Grand Tour and after (1660–1840)," in Peter Hulme and Tim Youngs, eds., *The Cambridge Companion to Travel Writing* (Cambridge: Cambridge University Press, 2002), 37–52.

7. James Simpson, *A Visit to Flanders, July, 1815* (Edinburgh: W. Blackwood, 1816), 54–56.

8. Seth William Stevenson, *Journal of a Tour through Part of France, Flanders, and Holland* (Norwich: Stevenson, Matchett, and Stevenson, 1817), 273–74.

9. Stevenson, *Journal*, 274–75.

10. John Booth published two separate maps of the battlefield in 1815 and 1816 that were attached to a popular history, but apparently also sold separately. *A Sketch of the Battle of Waterloo Fought Sunday June 18th, 1815* (London: Sold by J. Booth). The copy I have consulted appears in John Booth, *The Battle of Waterloo*, 4th ed. (London: J. Booth and T. Egerton, 1815).

11. See Daniel Boorstin, *The Image: A Guide to Pseudo-Events in America* (New York: Harper, 1961), 86–125; Withey, *Grand Tours and Cook's Tours*, especially vii–xii; Daniel Boorstin, Shelley Baranowski, and Ellen Furlough, eds., *Being Elsewhere: Tourism, Consumer Culture, and Identity in Modern Europe and North America* (Ann Arbor: University of Michigan Press, 2001); and Dean McCannell, *The Tourist: A New Theory of the Leisure Class* (New York: Schocken Books, 1976).

12. See Brendon Piers, *Thomas Cook: 150 Years of Popular Tourism* (London: Secker & Warburg, 1991); Rudy Koshar, ed., *Histories of Leisure* (Oxford and New York: Berg, 2002); Nicholas T. Parsons, *Worth the Detour: A History of the Guidebook* (Stroud, Gloucester, UK: Sutton, 2007); Jonathan Keates, *The Portable Paradise: Baedeker, Murray, and the Victorian Guidebook* (London: Notting Hill Editions, 2011); and Innes M. Keithren, Charles W. J. Withers, and Bill Bell, *Travels into Print: Exploration, Writing, and Publishing with John Murray, 1773–1859* (Chicago: University of Chicago Press, 2015).

13. Recent scholarship on the war includes: Alan Taylor, *The Civil War of 1812: American Citizens, British Subjects, Irish Rebels, & Indian Allies* (New York: Alfred A. Knopf, 2010); Troy Bickham, *The Weight of Vengeance: The United States, the British Empire, and the War of 1812* (New York: Oxford University Press, 2012); and J. C. A. Stagg, *The War of 1812: Conflict for a Continent* (Cambridge and New York: Cambridge University Press, 2012).

14. Martin Brückner, *The Social Life of Maps in America, 1750–1860* (Chapel Hill: University of North Carolina Press and the Omohundro Institute of Early American History and Culture, 2017); and Walter Ristow, *American Maps and Mapmakers: Commercial Cartography in the Nineteenth Century* (Detroit: Wayne State University Press, 1985).

15. Zadok Cramer, *The Navigator* (Pittsburgh: Cramer & Spear, 1821), 38–39.

16. For example (citing first editions, original titles, and publishers): Gideon M. Davison, *The Fashionable Tour* (Saratoga Springs, NY: G. M. Davison, 1822); Theodore Dwight, *The Northern Traveller* (New York: Wilder & Campbell, 1825); Horatio A. Parsons, *A Guide to Travelers Visiting the Niagara Falls* (Buffalo, NY: Oliver G. Steele, 1834); John Disturnell, *The Northern Traveller* (New York: John Disturnell, 1844); John Disturnell, *The Western Traveller* (New York: John Disturnell, 1844); and Orville Luther Holley, *Picturesque Tourist: Being a Guide through the Northern and Eastern States and Canada* (New York: John Disturnell, 1844).

17. It was destroyed, presumably by anti-British activists, in 1840, and rebuilt in 1853.

18. William Darby, *Tour from the City of New-York, to Detroit, in the Michigan Territory* (New York: Published for the Author by Kirk & Mercein, 1819), 169–70.

19. Darby, *Tour*, 185.

20. See Michael Twyman, "The Illustration Revolution," in D. McKitterick, ed., *The Cambridge History of the Book in Britain* (Cambridge: Cambridge University Press, 2009), 117–43. On its importance to guidebook publishing, see Keithren, Withers, and Bell, *Travels into Print*, 133–74.

21. For example, *The Illustrated London News* (founded 1842), *L'Illustration* (1844), *Illustrierte-zeitung* (1843), *Harper's* (1850), and *Leslie's Illustrated Newspaper* (1855).

22. The first (1832) edition of Karl Baedeker's guidebook to the Rhine included only one map. By the ninth (1856) edition, this had expanded to nine regional maps and ten city plans.

23. Benson J. Lossing, *The Pictorial Field-Book of the Revolution*, 2 vols. (New York: Harper & Bros., 1851–1852); and Lossing, *The Pictorial Field-Book of the War of 1812* (New York: Harper & Bros., 1868).

24. Lossing, *War of 1812*, unpaginated preface.

25. *Oxford English Dictionary* (online), www.oed.com, s.v. "field book."

26. For instance, see Lossing, *War of 1812*, 827–28.

27. Lossing, *War of 1812*, 54–55.

28. On landscape, time, and memory in Native American culture, see David Hurst Thomas, *Skull Wars: Kennewick Man, Archaeology, and the Battle for Native American Identity* (New York: Basic Books, 2000); Peter Nabokov, *A Forest of Time: American Indian Ways of History* (Cambridge: Cambridge University Press, 2002); and Adrienne Mayor, *Fossil Legends of the First Americans* (Princeton, NJ: Princeton University Press, 2005).

29. Megan A. Conrad, "From Tragedy to Tourism: The Battle of Gettysburg and Consumerism," MA thesis, Pennsylvania State University at Harrisburg (Dec. 2015), online at https://scholarsphere.psu.edu/downloads/5qf85n935d.

30. Lloyd, *Battlefield Tourism*, 19–23.

31. William F. Thompson, *The Pictorial Reporting of the American Civil War* (1959; repr., Baton Rouge: Louisiana State University Press, 1994); Ephraim George Squier, *Frank Leslie's Pictorial History of the American Civil War* (New York: F. Leslie, 1862); and Alfred H. Guernsey, *Harper's Pictorial History of the Great Rebellion*, 2 vols. (Chicago: McDonnell Bros., 1866–1868).

32. David C. Bosse, *Civil War Newspaper Maps: A Historical Atlas* (Baltimore: Johns Hopkins University Press, 1993).

33. Harold Holzer and Mark E. Neely Jr., *Mine Eyes Have Seen the Glory: The Civil War in Art* (New York: Orion, 1993), 171–204.

34. Quoted in *Panorama of the Battle of Gettysburg Permanently Located at Cor. Wabash Avenue and Hubbard Court* (Chicago, 1883?).

35. Boston: John B. Bachelder, 1873, and other dates of publication.

36. Richard Sauers, "Introduction," in *The Bachelder Papers: Gettysburg in Their Own Words*, ed. David L. Ladd and Audrey L. Ladd, 3 vols. (Dayton, OH: Morningside House for the New Hampshire Historical Society, 1994–1995), vol. 1, 9–14.

37. *Gettysburg Battlefield. Battle Fought at Gettysburg, Pa., July 1st, 2d & 3d, 1863 by the Federal and Confederate Armies.* Cited in Sauers, "Introduction," 10.

38. Meanwhile, he expanded his interviews to include Confederate commanders after the war. This allowed him to sketch a depiction of the climactic moment in the three-day battle, Major General George Pickett's famous, futile uphill charge on the Union center on Cemetery Ridge. In 1870 he commissioned a large-scale panoramic painting by James Walker, with which he toured the country.

39. Bachelder, "Introduction," 14–15.

40. The railroad's park was closed in 1896 by the Gettysburg Memorial Association, which had acquired the land.

41. Thomas E. Jenkins, *Gettysburg in War and Peace* (Baltimore: Press of J. Cox's Sons for Western Maryland Railroad Company, 1890).

42. Conrad, "From Tragedy to Tourism," 24–29.

43. William H. Whitney, *Union and Confederate Campaigns in the lower Shenandoah Valley Illustrated: Twenty years after* (Boston: W. H. Whitney, 1883).

44. Whitney, "Battlefield of Winchester, Va.," in *Union and Confederate Campaigns in the Lower Shenandoah Valley Illustrated*, map 33.

45. John Tregaskis, *The Battlefield of Gettysburg: The Men Who Fought Here and the Monuments Dedicated* (New York: Tregaskis & Co., 1888).

46. Grand Army of the Republic, *Manual for Annual Encampments Issued by the Passenger Department, Chesapeake & Ohio Ry.* (Chicago?, 1906).

47. Some of these guidebooks were actually published during the war, primarily for propaganda purposes. See Stephen Harp, *Marketing Michelin: Advertising and Cultural Identity in Twentieth-Century France* (Baltimore: John Hopkins University Press, 2001), 89–125. For a longer treatment of the use of maps and illustrations in these guides, see James R. Akerman, "Mapping, Battlefield Guidebooks, and Remembering the Great War," in Elri Liebenberg, Imre Josef Demhardt, and Soetkin Vervust, *History of Military Cartography: 5th International Symposium of the ICA Commission on the History of Cartography* (Berlin: Springer Verlag, 2016), 159–77.

48. Lloyd, *Battlefield Tourism*, 100–103.

49. *Michelin guide to the battlefields of the World War, vol. 1, the first battle of the Marne, including the operations on the Ourcq, in the marshes of St. Gond and in the Revigny Pass, 1914* (Milltown, NJ: Michelin, 1919), 120–21. I discuss this excursion at greater length in "Mapping, Battlefield Guidebooks, and Remembering the Great War."

50. Harry S. Howland and James A. Moss, *America in Battle* (Paris: Herbert Clarke, 1927), 492–93.

51. *A Guide to the American Battle Fields in Europe* (Washington, DC: Government Printing Office for the American Battle Monuments Commission, 1927).

52. Shaffer, *See America First*; Susan Sessions Rugh, *Are We There Yet? The Golden Age of American Family Vacations* (Lawrence: University of Kansas Press, 2008); and Akerman, "Twentieth-Century American Road Maps and the Making of a National Motorized Space."

53. Tennessee, North Carolina, West Virginia, and Virginia have published additional series.

54. Terry Copp [et al.], *1812: A Guide to the War and Its Legacy* (Waterloo, Ont.: LMSDS Press of Wilfrid Laurier University, 2013).

55. Copp et al., *1812*, 210.

CONTRIBUTORS

JAMES R. AKERMAN is director of the Hermon Dunlap Smith Center for the History of Cartography and curator of maps at the Newberry Library in Chicago. His research and publications primarily concern the history of transportation and tourist mapping, popular cartography, atlases, and the use of historical maps in education. He is the editor or coeditor of five books: *Cartography and Statecraft* (1999); *Cartographies of Travel and Navigation* (2006); with Robert W. Karrow, *Maps: Finding Our Place in the World* (2007); *The Imperial Map* (2009); and *Decolonizing the Map* (2017).

VERONICA DELLA DORA is professor of human geography at Royal Holloway, University of London. Her research interests and publications span historical and cultural geography, the history of cartography, and Byzantine studies with a specific focus on landscape, sacred space, and the geographical imagination. She is the author of three books: *Imagining Mount Athos: Visions of a Holy Place, from Homer to World War II* (2011); *Landscape, Nature, and the Sacred in Byzantium* (2016); and *Mountain: Nature and Culture* (2016).

BARBARA E. MUNDY is professor of art history at Fordham University. She studies the art and visual culture produced in Spain's colonies, with particular emphasis on indigenous cartography. Her scholarship spans both digital and traditional formats. She is the author of two books: *The Mapping of New Spain: Indigenous Cartography and the Maps of the Relaciones Geográficas* (1996); and *The Death of Aztec Tenochtitlan, the Life of Mexico City* (2015). With Dana Leibsohn, she is the creator of the online resource *Vistas: Visual Culture in Spanish America, 1520–1820*.

RICHARD A. PEGG is director and curator of Asian art for the MacLean Collection of Illinois. He is the author of *A Walk through the Ages: Chinese Archaic Art from the Sondra Landy Gross Collection* (2004); *Passion for Form: Selections of Southeast Asian Art from the MacLean Collection* (2007); *The MacLean Collection: Chinese Ritual Bronzes* (2010); and *Cartographic Traditions in East Asian Maps* (2014).

WILLIAM RANKIN is associate professor of the history of science at Yale University. He is the author of *After the Map: Cartography, Navigation, and the Transformation of Territory in the Twentieth Century* (2016). Also a cartographer, he has published and exhibited his own maps in the United States and Europe. He is currently working on a new book, *Radical Cartography: Visual Argument in the Age of Data*, under contract with Viking Books.

DANIEL ROSENBERG is professor of history at the University of Oregon. He is an intellectual and cultural historian, with special interests in the history of information graphics. He has published two books: with Susan Harding, he edited *Histories of the Future* (2005), and with Anthony Grafton, he is the author of *Cartographies of Time: A History of the Timeline* (2010). Rosenberg is also editor-at-large of *Cabinet: A Quarterly of Art and Culture*, where he is a frequent contributor. His current research concerns the history of data.

SUSAN SCHULTEN is professor of history at the University of Denver. Her area of specialty is nineteenth- and twentieth-century US history, and the history of maps. She is the author of four books: *The Geographical Imagination in America, 1880–1950* (2001); *Mapping the Nation: History and Cartography in Nineteenth-Century America* (2012), *A History of America in 100 Maps* (2018); and, with Elliott Gorn and Randy Roberts, *Constructing the American Past: A Sourcebook of a People's History* (2018).

ABBY SMITH RUMSEY is a historian of ideas, focusing on how information technologies shape perceptions of the past and of personal and cultural identity. She received a PhD from Harvard and has taught at both Harvard and Johns Hopkins University. She has also overseen the Scholarly Communication Institute at the University of Virginia, directed programs at the Council on Library and Information Resources, held multiple roles at the Library of Congress, and spearheaded the National Science Foundation's Blue Ribbon Task Force report on "Sustainable Digital Preservation and Access." The author of *When We Are No More: How Digital Memory Is Shaping Our Future* (2016), she has written and lectured widely on digital preservation, online scholarship, the nature of evidence, the changing roles of libraries and archives, intellectual property in the digital age, and the impact of new information technologies on perceptions of history and time.

KÄREN WIGEN is Frances and Charles Field Professor of History at Stanford University. Her areas of expertise are early modern Japanese history and the history of cartography. She has published four books: *The Making of a Japanese Periphery, 1750–1920* (1995); with Martin Lewis, *The Myth of Continents: A Critique of Metageography* (1997); *A Malleable Map: Geographies of Restoration in Central Japan, 1600–1912* (2010); and, coedited with Sugimoto Fumiko and Cary Karakas, *Cartographic Japan: A History in Maps* (2016). She is currently working with Martin Lewis on a book about the misleading cartographies of sovereignty.

CAROLINE WINTERER is William Robertson Coe Professor of History and American Studies at Stanford University. An intellectual historian, she is the author of four books: *The Culture of Classicism: Ancient Greece and Rome in American Intellectual Life, 1780–1910* (2002); *The Mirror of Antiquity: American Women and the Classical Tradition, 1750–1900* (2007); *The American Enlightenment: Treasures from the Stanford University Libraries* (2011); and *American Enlightenments: Pursuing Happiness in the Age of Reason* (2016). She is currently writing a book on the history of deep time.

INDEX

Page numbers in italics refer to figures.

"Aboriginal Wandering" (Willard), 175, 189
Accademia degli Argonauti, 116
Adams, Herbert Baxter, 187
Agassiz, Louis, 161, 164–66, 170n52
Aizu-Wakamatsu, 58
Akerman, James R., 12, 147–48, 193–218, 219
Aleni, Giulio, 41
Alexander VII Chigi (pope), 109
allegory: Japan and, 56, 58; language and, 135–36, 145; veils and, 106–12, 115, 118, 121n1, 123n28, 123n39
Allegory with Venus and Cupid, An (Bronzino), 107–8
Allen, Joel, 187
All under heaven map of the everlasting unified Qing empire (Huang Qianren), 70
Along the Trail with Lewis and Clark (Fifer and Soderberg), 194
America, 1; *Atlas historique* and, 114; deep time and, 149–67; Native Americans and, 155, *156*, 175, 183, 187, 202, 217n28; as New World, 11, 147, 161, 165; Revolutionary War and, 164
America in Battle (Howland and Moss), 210
American Association for the Advancement of Science, 177
American Battle Monuments Commission, 210–11
American Civil War, 12, 23, 30; Antietam and, 202, 207; apps for, 213; Bachelder on, 204–6, 212, 214; brochures on, 212–13; capture of New Orleans and, 179; Chickamauga and, 207; Confederacy and, *180*, 181, 203, 205, 206, 209, 212, 217n38; early US time maps and, 178–83, 188; effects of, 202–3; Fredericksburg and, 202; Gettysburg and, 147–48, 179, 181, 202–8, 212; Lincoln and, 180–81, 202; Lindenkohl map and, 179–81; monuments and, 207–8; sesquicentennial of, 212; Shiloh and, 202; tourist maps for, 196, 202–9, 212–13; Union and, 179–82, 188–89, 203–9, 212, 217n38; US Coast Survey and, 178–82; Vicksburg and, 179, 181
American Geographical Society, 16
"American Panorama" project, 190–91
Amerindians. *See specific tribe*
"Analytical Language of John Wilkins, The" (Borges), 129–30
Anderson, Benedict, 195
Anglicanism, 123n30, 133
animation, 189; collapsed, 17, 24, 27, 29, 177; computer, 16, 18; digital, 5, 18; flow and, 4, 17–18, 26, 29; Geographic Information Systems (GIS) and, x, 4–5; historical, 3–4, 27, 29–30; linear succession and, 26; snapshots and, 3–4, 17, 24, 26; zoopraxiscope and, 33n21
Annals of the Joseon dynasty, 71
Antietam, 202, 207
antiquarianism, 105, 112
Army Corps of Engineers, 15, 181
Arrowsmith, Aaron, 69
Art as Experience (Dewey), 4
Association of American Geologists and Naturalists, 155
astronomy, 70–71, 80, 110, 134, 136
Atlante veneto (Coronelli), 116–20
Atlas geográfico (Garcia Cubas), 98
Atlas historique (Châtelain), 112–14
Atlas maior (Blaeu), 116
Atlas of Eternal Peace, 47
Atlas of the Historical Geography of the United States (Jameson), 22–23, 28, 171, 182, 187–91
Austin, Gilbert, 138, *139*

Austria-Hungary, 23
Aztecs, ix; afterlife beliefs of, 98–100; Aztlan and, *xii*, 1, 84–86; calendars and, 79–85, 91, 93, 97, 100, 101nn13–14, 150, 154; Chapultepec and, 85, 89; Codex Mendoza and, 91–97; conquest and, 89–91; creation stories of, 81–84; deep time and, 97, 100, 150, 154; Ehecatl and, 81, 91; Elias on, 79, 82; Huitzilopochtli and, 86; human sacrifice and, 81, 91, 101n12; *Hystoyre du Mechique* and, 81–84, 89, 93, 95, 97; iconography and, 10, 91, 95, 97; Mapa de Sigüenza and, *xii*, 85, 89, 94–95, 98–100; maps of movement and, *xii*, 1, 81, 84–85, 97, 101n11; Mexico and, 80–89, 93–94, *96*, 98–100; migration and, 84–89; Nahua culture and, 81–85, 89–90, 93, 97, 101n11; Nanahuatzin and, 81; narrative style of, 94–98; regime of historicity and, 80; solar years and, 82, 84–85; space/time beliefs of, 1, 12, 81–84, 94–98; Spanish conquest and, 77, 80, 93, 97, 100; Tenochtitlan and, 1, 6–7, 10, 84–86, 89, 92, 93–97, 100; Tepanecs and, 85; Tezcatlipoca and, 81–82, 91; Tlaltecuhtli (Earth Lord) and, 81, 95, 97; Tlatelolco and, 89; violence of, 81–82, 90–91; world-making and, 84–91; Zolipa map and, 97–98
Aztlan, *xii*, 1, 84–86

Bache, Alexander, 178–79
Bachelder, John B., 204–6, 212, 214
Bacon, Albert, 138
Bacon, Francis, 110, 112
Baedeker, Karl, 198, 200
banners, 70, 104, 115–16, 124n61
Baroque style, 104–5, 108–9, 120
Barthes, Roland, 140–42

[221]

Bashō, 49
Basil of Caesarea, 104
Batoni, Pompeo, 108
Bender, John, 140–41
Bergson, Henri, 26–27
Bernini, Gian Lorenzo, 108–9
Bible, 107; chronology and, 11; deep time and, 100, 149–50, 154, 165–67; Dutch and, 112, 122n13; Medieval Christianity and, 38; Ortelius and, 39; universal narrative and, 114
Black, Jeremy, 37–39
Blaeu, Joan, 116
Bodenhamer, David, 16
Boke of Idrography (Rotz), 136, 137
Book of Sand, The (Borges), 167
Boone, Elizabeth, 95
Booth, John, 216n10
Borges, Jorge Luis, 129–31, 167
Bradley, Abraham, Jr., 172
British East India Company, 69
British Library, xi
Brock, Isaac, 199
Bronzino, Agnolo, 107–8
bubble grid technique, 30
Buddhist maps, 38–39
Buohl's Illustrated Map and Guide to Tour the Gettysburg Battlefield, 212

Caesar, Julius, 114
calendars, 2, 35; Aztecs and, 79–85, 91, 93, 97, 100, 101nn13–14, 150, 154; Chinese, 71, 75n28; Gregorian, 25, 80; Julian, 79–80; Korean, 71; leap days and, 80; precision in, 79–80; solar years and, 82, 84–85; Tovar, 82, 83, 93
Cambrian period, 170n50
Capital edition of the complete map [based on] astronomy (Ma Junliang), 70
Carnegie Institution, 188
Cartesianism, 4, 13n14, 63, 79
cartographic moment, 18–25
cartographic shutter, 18, 22, 24, 26, 31

cartouches, 10, 41, 54, 104, 116
Casaubon, Isaac, 110–11
Casey, Edward, 79
Casilear, John W., 161
Catholicism, 10, 65, 80, 100, 108, 115, 118. See also Jesuit maps
Cemetery Ridge, 212, 217n38
censorship, 35, 54–59
census data, 11, 18–19, 23, 30, 182–90
Chambersburg Pike, 212
Chapultepec, 85, 89
Chardin, Jean, 115
Chart of the World on Mercator's Projection (Arrowsmith), 69
charts: Coast Survey and, 178; genealogical, 53; Jesuit maps and, 69, 71; postal maps and, 172, 173; scale and, 19; various information on, 188; Wilkins and, 131
Châtelain, Zacharias, 112–14
Chen Lunjiong, 69
Cheonhado (world maps), 72
Chesapeake & Ohio railroad, 208–9
Chicago's World Columbian Exposition, 203
Chicago Tribune, 203–4
Chichimeca Nonohualca, 89–90
Chickamauga, 207
Chikuma River, 55
China, 2, 9; Atlas historique and, 114; Buddhist maps and, 38–39; calendars and, 71, 75n28; Chongzhen calendar and, 75n28; gujin and, 63–65, 69–73; Japanese maps of, 42–45, 47; Jesuit maps and, 35, 41, 63–73, 75n28; literate class and, 35; Ma Junliang and, 70–71; Ming dynasty and, 9, 41, 43, 63–74, 75n25, 75n28; natural world order (zhi) and, 66; old cartographic tradition of, 42–43, 48; Qianlong era and, 64, 68, 69–70, 74n10; Qing dynasty and, 35, 43, 45, 63–65, 70–71, 74–75; Sancai Tuhi and, 43, 68–70, 72,

75n18, 75n24; Song dynasty and, 6, 42, 44; time maps and, 45; Wang Qi and, 43, 68–70, 72, 75n18, 75n24; Western trade and, 76, 76n31; woodblock maps and, 65, 67–73, 74n1; Zhang map and, 67–68; Zhou dynasty and, 43; Zhuang and, 63–66, 69–71
China Illustrata (Kircher), 124n52
Chippawa, 199
Chironomia, or, a Treatise on Rhetorical Delivery (Austin), 138, 139
Choice of Emblems (Whitney), 106
Chongzhen calendar, 75n28
choropleth maps, 34n37
Christians, 38, 100, 107, 149–50
chronophotography, 26
Clavigero, Francisco Javier, 98
Clio, 111
Codex Mendoza, 91–97
Codex Nuttall, 102n25
collapsed animation, 17, 24, 27, 29, 177
colored lines, 179, 188
Columbus, Christopher, 6, 175
Complete geographical map of mountains and streams (Zhang Hong), 69
Complete Map (Ricci), 65
Complete map of ten thousand countries of the world (Jesuit map), 65, 66
compression, ix–x, 5
computers: animation and, 16, 18; GIS and, 4, 13n14; hypertext and, 131; language and, 131, 144n23; projection of space and, 138
conceptual maps, 78, 127, 130, 133
Confederacy, 180, 181, 203, 205, 206, 209, 212, 217n38
Confucianism, 45, 66
consciousness, 26–27, 148, 159–60, 167
Cook, James, 69
Cook, Thomas, 198
Coronelli, Vincenzo, 105, 116–20, 124n61
Corps of Discovery, 194
Cosgrove, Denis, 116

Craan, William Benjamin, 197–98
Cramer, Zadok, 199
Creation, 81–84, 98, 100, 116, 154–55, 165
Crimean War, 200
Critique of Pure Reason (Kant), 158–59
cross-sections, 160
Cuauhtinchan, 90–91
Culhuacan, 94
Culp's Hill, 212
current events map, 54–60
Currier and Ives, 164

daily records of the royal secretariat, The, 71
d'Alembert, Jean le Rond, 139, 140
Dante, 106
Darby, William, 199–200
Darwin, Charles, 158–59, 165
data visualization, x, 172
David Rumsey Map Center, xiii, 1–2, 113, 117, 142, 152–53, 156, 174, 190
deep time: Aztecs and, 97, 100; calendars and, 150, 154; Cambrian period and, 170n50; concept of, 11, 149; cross-sections and, 160; diagrams and, 150, 157, 160–61; discovery and, 155, 161; earth strata and, 151–57; first American maps of, 149–67; flow and, 149, 154, 157, 160–61; fossils and, 149, 151, 153, 155, 157, 160, 164; geological, 147, 149–67, 170n50; human brain and, 157–61; landscape painting and, 161–67; Lyell and, 151, 155, 158; Maclure and, 151–52; maps of movement and, 161; New World and, 11, 147, 161, 165; Old World and, 11, 165; Owen and, 151, 153–57, 168n8; paintings and, 150, 155–57, 161, 167, 170n52; Silurian period and, 164–65, 170n50; temporality and, 167; trilobites and, 150, 164–66; Willard and, 153–55
Degérando, Joseph-Marie, 127–28, 131
Delafield, Richard, 181
Della Dora, Veronica, 10, 77, 103–25,

144n29, 219
Des Moines River, 155–57
Dewey, John, 4
diagrams: deep time and, 150, 157, 160–61; grammar and, 2, 11, 134–36, 140, 145n33; humanist studies and, 1–2; language and, 131–43, 145n34; T-O, 38; war and, 206; Wilkins and, 131–43, 145n34
Dickens, Charles, 104
Diderot, Denis, 139, 140
Digital Scholarship Lab (DSL), 190–91
digital technology: advantages of, 2–3, 5, 12; exploring old maps by, 5; GIS and, 2 (*see also* Geographic Information Systems [GIS]); humanists and, 25, 172, 188; Japan and, 52; preservation of historical knowledge and, x–xi; Republic of Letters project and, 52; spatial representation and, 16, 25; temporality and, 5, 9, 12, 17, 25, 27, 148; United States and, 148, 172, 188–91
discovery: deep time and, 155, 161; European voyages of, 111, 175; importance of, 105; maps of, 177–79, 188–89, 192n12; tension between known and unknown and, 110; US maps and, 177–78; veils and, 104–5, 110–16, 120–21
diversity, 7, 12, 105, 147
Drake, Francis, 179
Durand, Asher B., 161
durational time, 3, 5, 10, 13n10, 19, 21, 25–26, 33n18
Dutch, 41, 105, 108, 111–12, 114, 116, 120, 122n13, 123n27
Dutch East India Company, 111
dynamism: animation and, 18 (*see also* animation); Coronelli and, 105; Geographic Information Systems (GIS) and, 3, 8; Hilgard and, 183; maps of movement and, 16 (*see also* maps of movement); rate over time and, 28; rivers and, 15; temporality and, 3, 8,

16, 18–19, 25–26, 73, 183; veils and, 105, 110, 120. *See also* static images

early modern age, 8; cartographic metaphors and, 10, 104, 114; East Asia and, 10, 35; Europe and, 78–79, 104, 114; grammatical diagrams and, 2, 11, 134–36, 140, 145n33; historical maps and, 5–6; Japan and, 9, 37–60, 220
East Asia, 9–10, 35. *See also specific country*
Eco, Umberto, 131
Edo Bay, 49–51
Edo period, 10, 42, 49, 51
Egypt, 19, 20, 25, 38
Ehecatl, 81, 91
1862 Antietam Campaign: Lee Invades Maryland (brochure), 212–13
1812: A Guide to the War and Its Legacy (guidebook), 214
Einstein, Albert, 26, 33n30
Elias, Norbert, 79, 82
Elizabeth I, 108
Encyclopédie (Diderot and d'Alembert), 139, 140–41
Epea Pteroenta; or The Diversions of Purley (Tooke), 132–33
Epoch Maps Illustrating American History (Hart), 184
Erasmus, 106–7
Essay Towards a Real Character, An (Wilkins), 126, 127–32, 133, 136, 141
Ethington, Philip, 26–27
event maps, 42, 58, 179
evolution, 158–61, 171

Fallen Timbers, 201
Father Time (Chronos), 10, 104–8, 110–12, 114, 120–21, 122n19
Federal Highway Administration, 212
Feeney, Denis, 4
Fifer, Barbara, 194
Figurative Map of the successive losses in

men of the French Army in the Russian campaign 1812–13 (Minard), x
First Battle of the Marne, 210
Fisk, Harold, 15, 17, 29–30
flow, 104; animation and, 4, 17–18, 26, 29; background, 4; calendars and, 82; capital, 17; cartographic moment and, 18–25; daily life and, 4; deep time and, 149, 154, 157, 160–61; GIS and, 3; information and, 172, 176; of language, 134; lived experience and, 79; maps of movement and, 3, 19, 21, 23, 26, 31, 161; simple lines and, 26–27; spatial, 17, 25, 27–28; succession and, 30; temporal, 3, 17, 19, 21, 24–31
flow lines, 21, 26–27, 37
Fludd, Robert, 136
Fort Donelson, 179
Fort Erie, 199
Fort George, 199
For the Day Present, or the Last Day (Withers), 103
Fort Henry, 179
fossils, 149, 151, 153, 155, 157, 160, 164
Foucault, Michel, 78, 127, 130, 136
Fox, Dixon Ryan, 192n22
Franklin, Benjamin, 52
Fredericksburg, 202
frontiers, 11, 98, 147–48, 172, 184–90, 199
frontispieces, 124n52; Aztec maps and, 93; Father Time and, 104, 110, 120; veils and, 104–5, 109–15, 120–21, 123n33, 123n35

Gadsden Purchase, 98
Gallatin, Albert, 189, 191n6
García Cubas, Antonio, 98, 100
Gellius, Aulus, 106
genealogy, 10, 12, 48, 53, 112, 179
Genesis, Bible book of, 38
Geographic Information Systems (GIS), 2, 12n6, 13n8; animation and, x, 4–5; Cartesian logic of, 13n14; computers and, 4, 13n14; dynamism and, 3, 8; flow and, 3; Historical GIS, 16, 171, 189, *190*; as new form of representation, x; Newtonian time and, 4
geology: Agassiz and, 161, 164–66, 170n52; Association of American Geologists and Naturalists and, 155; change maps and, 30; creation and, 154; cross-sections and, 160; deep time and, 147, 149–67, 170n50; Fisk and, 15; fossils and, 149, 151, 153, 155, 157, 160, 164; Grand Canyon and, 161; layering and, 29; Lyell and, 151, 155, 158; Maclure and, 151–52; metaphor and, 121; Murchison and, 160, 164, 170n50; Owen and, 151, 153–57, 168n8; sedimentation and, 30; Silurian period and, 164–65, 170n50; strata and, 29, 38, 151–57, 159–60; trilobites and, 150, 164–66; Yosemite and, 161
geovisualization, 16, 32n8
Gettysburg, 147–48, 179, 181, 202–8, 212
Gettysburg Monument Commission of the State of New York, 208
Gettysburg: What to See, and How to See It (Bachelder), 204–5
Gilbert, Humphrey, 175
Gilman, Daniel Coit, 187, 188
globalization, 12, 74
globes, x, 104, 116
Glorieuse campagne de Monseigneur le duc d'Anguyen commandant les armées de Louis XIII (Pontault), 115
Goffart, Walter, 37–39, 114
Goodchild, Michael, 3, 16
Grafton, Anthony, 134
Graham, Charles R., 208
grammar, 2, 11, 134–36, 140, 144n20, 145nn33–34
Grand Army of the Republic, 206
Grand Canyon, 161
Grand Tour, 197, 216n6
graphic visualization, 32n8, 77; Aztecs and, 79–80, 82, 84; early US time maps and, 174–75, 189; Japan and, 37; language and, 133–34, 136, 138, 142–43; photo-cinematic idiom and, 23; temporalities and, 5; user-centered, 16
Great Depression, 171
Great Flood of the Mississippi River, 15
Great Hurricane of 1938, 30
Great Lakes, *185*, 199–200
Great Plains, 190
great Qing dynasty world map of tribute-bearing countries with spherical coordinates, then and now, The (Zhuang), 63, *64*
Great War, 209–14
Greeks, 38–39, 41, 110, 114, 118
Gregorian calendar, 25, 80
Gregory, Ian, 16
Guatemala, 84
Gueudeville, Nicolas, 111
guidebooks, 217n20, 217n22, 218n47; Lewis and Clark Expedition and, 193–95; war sites and, 196–214
Guides Joannes (French guidebooks), 200
Guide to the American Battle Fields in Europe, A (American Battle Monuments Commission), 210–11
gujin, 63–65, 69–73
Gulf of Mexico, *178*, 200

Hall of Charts and Maps, 178
Harper's (journal), 203
Hart, Albert Bushnell, 184, 192n22
Haseltine, William Stanley, 170n52
Hashimoto Kenjiro, 50. *See also* Utagawa Sadahide
Heade, Martin Johnson, 161–64, 166–67
Henry VIII, 108, 136
Hereford map, 38
Hilgard, Julius, 179, 183–84, 190
Hiraizumi, 50
Hiroshige, 56
Historia Tolteca-Chichimeca, 89–91

[224] INDEX

historical atlases, 28; diversity and, 7; early modern age and, 5–6; Japan and, 37, 39, 42–47; rise of genre of, 37; United States and, 11, 172, 174–75, 192n23; universal atlases and, 124n48; Vidal and, 19, 27; Wright on, 172
Historical GIS, 16, 171, 189, *190*
"Historical Sketch of the Rebellion" (Lindenkohl), 179–81, 183, 188–89
historical time, 1, 3–4, 27, 112, 114, 154
historical variable, 17, 27–31, 33n34
History of the United States (Willard), 174–75
Hiyama Gishin, *46*
Hokusai, Katsushika, *45*, 47
Holy Land, 38–39
Holy League, 118
Honshu, 49, 58
Hooghe, Romeyn de, 111
Hooke, Robert, 138, 146n45
Hornaday William Temple, *186*, 187
Howland, Harry S., 210
Huang Qianren, 70
Huang Zongxi, 70
Hudson River School, 161
Huitzilopochtli, 86
human brain, 157–61
humanists, 1, 25, 145n34, 172, 188
Humboldt, Alexander von, 158
hurricanes, 30, *31*, 32n17
hybrid organization, 141
hyperobject, 3
hypertext, 131
"Hystoyre du Mechique," 81–84, 89, 93, 95, 97

iconography, x; allegory and, 56, 58, 106–12, 115, 118, 123n28, 123n39, 135–36, 145; Aztecs and, 10, 91, 95, 97; biblical, 38–39; medieval, 38, 41, 106; T-O map and, 38; toponyms and, 6–7, 35, 39, 43, 47, 49, 65; veils and, 10, 106, 109–11; war and, 210

Iconologia (Ripa), 106
Illustrated compendium of the three powers (Sancai Tuhui) (Wang Qi), 43, 68–70, 72, 75n18, 75n24
Illustrated London News, 200
Illustrated Sino-Japanese Encyclopedia (Wakan Sansai Zue) (Terajima), 43
Illustrious Holland (Van Leeuwen), 111
immigrants, 21–22, 25, 179, 182–84
India, 23, 33n22, 38, *41*, 69, 111
insets, x, 58, 104, 172
Isis, 110, 123n33
isochronic maps, 24
isometric maps, 205
Israelites, 38–39

Jackson, J. Hughlings, 159
Jacob, Christian, 103
James, Edwin, 152, *153*
James, John Franklin, 171, 187–91
James, William, 159–61, 169n38
Jameson, John Franklin, 171, 187–90, 192n25
Japan, ix, 1–2; ancient period and, 48; *Atlas historique* and, 114; cartographic tradition of, 43; civil war and, 47–48, 51; current events maps and, 54–60; digital maps and, 52; early modern age and, 9, 37–60, 220; Edo period and, 10, 42, 49, 51; historical atlases and, 37, 39, 43–45, 47; Imperial loyalists and, 57–58; Jesuit maps and, 41; literate class in, 35; mapping news in, 54–60; maps of China in, 42–45, 47; medieval times and, 39, 41, 47–49, 51, 55, 61n4; native-place mapping in, 47–54; nativism in, 52–53; Perry and, 56–57; Restoration wars and, 58; time maps and, 45, 54; tourist maps and, 49, 58; woodblock maps and, 41–43, 49
Japanese Alps, 49, 54
Jesuit maps, 36; Aleni and, 41; charts and, 69, 71; China and, 35, 41, 63–73, 75n28; Clavigero and, 98; *Complete map of ten thousand countries of the world* and, 65, *66*; court surveys and, 75n21; East Asia and, 9–10; Japan and, 41; Korea and, 71–74; Ortelius and, 41; Ricci and, 9–10, 41, 43, 56, 58, 65–71, 74n12, 75n25, 75n28; spherical earth of, 66; static images and, 73
Jesus Christ, 38
John Carter Brown Library, xi
Johns Hopkins University, 187
Jordan River, 38–39
Joseon dynasty, 41, 71, 75n25, 75n28
Journal du Voyage du Chevalier Chardin (Chardin), 115
Julian calendar, 79–80

Kamakura, 58
Kant, Immanuel, 158–59
Katsushika Hokusai, *45*, 47
Kawanakajima, 55
kawaraban (broadsheets), 54–58
Kensett, John F., 161
Kepler, Johannes, 110
Key to Bachelder's Isometrical Drawing of the Gettysburg Battle-field, with a Brief Description of the Battle (Bachelder), 205
Kircher, Athanasius, 52, 124n52
Knights Hospitaller, 118
Knothe, Herbert, 16
Kobayashi Kōhō, 56
Kohl, Johann Georg, 177–79
Korea, 2; calendars and, 71; cartographic tradition of, 43; *Cheonhado* (world maps) and, 72; *Gangnido* map and, 71; Jesuit maps and, 71–74; Joseon dynasty and, 41, 71, 72–73, 75n25, 75n28; literate class and, 35; Sukjong and, 71; *Yeojido* (printed atlases) and, 72; Yi family and, 71
Kyoto, 52–53

Lairesse, Gerard de, 114
Lake George (Heade), 161–64, 166–67
landscape painting, 2, 11, 35, 60, 150, 155, 157, 161–67
language: Borges and, 129–31, 167; computers and, 131, 144n23; as conceptual map, 127–48; Degérando on, 127–28, 131; diagrams and, 131–43, 145n34; *An Essay Towards a Real Character* and, 126, 127–32, *133*, 136, 141; flow of, 134; Foucault and, 127, 130, 136; grammar and, 11, 134–35, 140, 144n20, 145n34; hybrid organization and, 141; maps of movement and, 138, 147; Mixtec, 89; Nahuatl, 81–85, 89–90, 93, 97, 101n11; prepositions and, 11, 78, 132–35, 138, 143, 144n26; time maps and, 134; veils and, 105; Wilkins and, 78, 126, 127–43, 144n13, 144n26, 145n34, 146n45
Laocoön (Lessing), 161
Last Leaves of American History (Willard), 176, *177*
layering, 29, 33n18, 159
Lee, Robert E., 213
legends, 19, 37, 136, *185*
Leslie's journal, 203
Lessing, Gotthold, 161
Leventhal Map Center, xi
Lewis and Clark Expedition, 193–95
Library of Congress, xi, 61n5, 62n24, 64, 74n1, *173*, 178, *180*, 183, 185–86, 192n12, 220
Life magazine, 24
L'Illustration (French newspaper), 200
Lincoln, Abraham, 180–81, 202
Lindenkohl, Henry, 179–83, 188
linear time, 4, 63
Little Round Top, 206, 212
Liu, Lydia, 132, 135, 144n23
lived experience, 4, 63, 79, 149
Li Zhizao, 65
Long, Stephen, 152, *153*
López Austin, Alfredo, 97

Lord of Mito, 57
Lossing, Benson J., 200–203, 205, 212, 214
Louis XIV, 116
Lundy's Lane, 199, 214
Lyell, Charles, 151, 155, 158

Macartney, George, 69
Maclure, William, 151–52
Ma Junliang, 70–71
Manchuria, 70
Manchus, 45, 70–72
Manual of Gesture (Bacon), 138
Mapa de Sigüenza, *xii*, 85, 89, 94–95, 98–100
Map of China (Huang Zongxi), 70
Map of the integrated regions and capitals of states over time (Korean map), 71
"Map of the Tracks of Yu" (*Yu ji tu*), 42
Map of the world (Ortelius), 67
Map of the world (Ricci), 67
Map of the World on a Globular Projection (Arrowsmith), 69
mappaemundi, 38–39
Map showing the location of battle fields of Virginia compiled from official records and maps (Chesepeake & Ohio), 208–9
maps of movement: Aztecs and, *xii*, 1, 81, 84–85, 97, 101n11; deep time and, 161; dynamism and, 16, 25; flow and, 3, 19, 21, 23, 26, 31, 161; language and, 138, 147; spatial change and, 26; time and, 3, 7, 10, 16, 19, 21–22, 24, 26, 31, 65, 81, 84–85, 93, 97, 138, 147, 161, 172–73, *175*, 183, 190, 193, 214–15; veils and, 120; war and, 193, 202, 208, 215
Marey, Étienne Jules, 24, 26–27
Marrinan, Michael, 140–41
Marx, Karl, 158
Maryland Civil War Trails mobile app, 213
Massachusetts Institute of Technology (MIT), 24, 182
Mathematical Magick (Wilkins), 138, *139*

mathematics, 4, 25–26, 42, 80, 110
McPhee, John, 167
Meade, George, 205
medieval era: Japan and, 39, 41, 47–49, 51, 55, 61n4; *mappaemundi* and, 38–39; T-O maps and, 38; triptychs and, 161
Memorie istoriografiche (Coronelli), 118, *119*
memory: associations and, 65; group, 8; lived experience and, 63; meteorological, 31; national, 12, 194, 215, 217n28; personal, 12; personification of, 106; political, 29; situated, 25; spatial, 12, 17–30; temporality and, 27, 29–30; urban, 29
Mendoza, Antonio de, 93
Mental Evolution in Man (Romanes), 160
Me-sa-sa, 201–2
Mesoamerica, 1–2, 10, 84, 89, 91, 95, 97
metaphor, 27, 31, 125n64; allegory and, 56, 58, 106–12, 115, 118, 123n28, 123n39, 135–36, 145; Baroque style and, 104–5, 108–9, 120; Coronelli and, 105, 116–20, 124n61; Dutch and, 105, 108, 111–12, 114, 116, 120, 122n13, 123n27; early modern age and, 10, 104, 114; geology and, 121; Nature and, 110; paintings and, 105, 108, 112, 123n28, 150, 155; photographic, 17–18; psychology and, 159–60; telescopic approach and, 112, 114, 120; Truth and, 105–10, 112, 114, 120, 123n28; veils and, 104 (*see also* veils); Venice and, 10, 105, 116, 118, *119*–20; Willard and, 155, 174
Mexican-American War, 98
Mexico: Aztecs and, 80–89, 93–94, 96, 98–100; Codex Mendoza and, 91–97; decline of bison and, 187; Gadsden Purchase and, 98; Gulf of, *178*, 200; independence of, 77, 80; *Last Leaves of American History* and, 176, *177*; Mapa de Sigüenza and, *xii*, 85, 89, 94–95, 98–100; Misantla region of, 95–97; regime of historicity and, 80; Spanish

conquest and, 77, 80, 93, 97, 100; Tenochtitlan and, 1, 6–7, 10, 84–86, 89, 92, 93–97, 100; Toltecs and, 89–91, 94, 97; Treaty of Guadalupe Hidalgo and, 177–78; Treaty of La Mesilla and Guadalupe and, 98; Valley of, 1, 10, 85, 88, 89, 94; Veracruz, 95; Zolipa map and, 97–98
Mexico City, 84, 86, 93, 100
Michelin guides, 209–10
Micrographia: or some Physiological Descriptions of Minute Bodies made by Magnifying Glasses (Hooke), 138, 146n45
microscopes, 109–10, 123n33, 138
migration, 7, 10, 17, 21–25, 84–89, 91, 93–94, 174, 175, 183–84, 187, 189
Mili, Gjon, 24
Mill, John Stuart, 158
Minard, Charles-Joseph, x, 23, 33n22
Ming dynasty, 9, 41, 43, 63–74, 75n25, 75n28
Mississippi River, 14, 15, 17, 29, 179, 181, 192n13, 194, 196, 199
Mito domain, 57
Mixtec language, 89
mnemomic devices, 145n34, 174
Mnemosyne, 106, 115, 120
Mongolia, 45, 56, 68–70, 114
Monmonier, Mark, 16
monuments, 91, 109, 115, 197, 199, 207–8, 210
More, Thomas, 53
Mori Yukiyasu, 51–52
Moss, James A., 210
Motoori Norinaga, 53–54
Mount Asama, 55–56
Mount Fuji, 58
Mount Tsukuba, 57
Moxey, Keith, 4
Mundy, Barbara E., 10, 12n1, 77, 79–102, 146n51, 219
Murchison, Roderick, 160, 164, 170n50

Murray, Edward, 198, 200
Mussulman, Joseph, 194
Muybridge, Eadweard, 23–24, 33n21

Nagakubo Sekisui, 43–45
Nagasaki, 42
Nahant, 170n52
Nahua culture, 81–85, 89–90, 93, 97, 101n11
Nanahuatzin, 81
Napoleon, 21, 197, 202
Narrative and Critical History of America (Winsor), 187
Native Americans, 155, 156, 175, 183, 187, 202, 217n28
nativism, 51–53
NATO, 77
neo-Marxism, 16
Neptune, 111
Newtonian time, 4, 77
New World, 11, 147, 161, 165
Niagara Falls, 196, 199, 214
Ninth Census, 182, 184
Northern Pacific Railroad, 193
nouveau theatre du monde, Le (Gueudeville), 111
Novum Organum (Bacon), 110

Ogden, C. K., 132–36, 144n23
Ohio River, 199
Old World, 11, 165
One Hundred Famous Views in the Various Provinces (Hiroshige), 56
Order of Things (Foucault), 130
orientation, 59, 78–81, 88, 89, 111, 124n61
Origin of Species, The (Darwin), 158
Ortelius, Abraham, 39, 41–42, 65, 67, 103, 123n41, 192n13
Osaka, 42, 51
Osaka Castle, 55
Osher Map Library, xi
Ottoman-Venetian War of the Morea, 118
Owen, David Dale, 151, 153–57, 168n8
Owen, Robert, 151

paintings: allegory and, 123n28; deep time and, 150, 155–57, 161, 167, 170n52; Father Time and, 105, 112; landscape, 2, 11, 35, 60, 150, 155, 157, 161–67; metaphor and, 105, 108, 112, 123n28, 150, 155; space and, 2, 35, 145n39, 161; war and, 203, 217n38
Panofsky, Erwin, 107–9
panoramas, 45, 47, 58, 60, 164, 176, 190, 203–5, 210, 212, 217n38
Parergon (Ortelius), 39, 41, 192n13
Paullin, Charles, 22, 33n24, 171, 188–90
Pax Tokugawa, 35
Pegg, Richard A., 9, 35, 63–78, 219
peregrination maps, 84, 86, 100
Perry, Matthew, 56–57, 200
Pettegree, Andrew, 122n13
Peutinger Table, 41
Phillippoteaux, Paul, 203
philosophy of time, 17, 25
photo-cinematic idiom: chronophotography and, 26; concept of, 17; dynamism of, 26; flow and, 22, 24–25; historical variable and, 27–31, 28; Marey and, 24, 26–27; Muybridge and, 23–24, 33n21; new culture of vision/perception and, 17; philosophy of, 25–27; succession of time and, 25; visual analysis of, 17
photography: cartographic shutter and, 18, 22, 24, 26, 31; chronophotography and, 26; defining cartographic moment and, 18–25; electronic flash and, 24; historical variable and, 27–31; new culture of vision/perception and, 17; philosophy of, 25–27; thematic mapping and, 18
Pickett's Charge, 203, 206, 212
Pictorial Field-Book of the Revolutions (Lossing), 200
Pictorial Field-Book of the War of 1812 (Lossing), 200, 201

Index [227]

"Picture of Nations" (Willard), 175–76
Pindar, 106
Plutarch, 106
Pontault, Sébastien, 115
population: census and, 11, 18–19, 23, 28–29, 30, 182–90, 192n17; Hilgard map and, 179, 183–84, 190; historical variables and, 28, 30; war and, 202
population maps, 182–90, 192n22
Poseidon, 111
postal mapping, 172, 173, 182
Postal Service Act, 172
Powell, John Wesley, 189
prepositions, 11, 78, 132–35, 138, 143, 144n26
Preston, John S., 121
Principles of Geology (Lyell), 155, 158
Principles of Psychology (Spencer), 158–59
Principles of Psychology, The (James), 159–61, 169n38
print maps, 11, 49, 70
print media, 2, 5, 10–11, 105
Promised Land, 38
propaganda, 98, 105, 218n47
Protestantism, 108
Prussia, 158
psychology, 149–50, 157–61

Qianlong era, 64, 68, 69–70, 74n10
Qing dynasty, 35, 43, 45, 63–65, 70–71, 74–75
Queenston Heights, 199

Ramirez, José Fernando, 100
Rankin, William, 7–9, 15–36, 73, 177, 183, 219
Ray, John, 129
Reformation, 108
Renaissance, 5, 105–6, 136
Report of a Geological Survey of Wisconsin, Iowa, and Minnesota (Owen), 155
Report of the Ninth Census (Walker), 182
Republic of Letters project, 52

Rethinking the Power of Maps (Wood), 37
Revolutionary War, 164
Ricci, Matteo, 9–10, 41, 43, 56, 58, 65–71, 74n12, 75n25
Richard, I. A., 132–33
Ripa, Cesare, 106
Romanes, George, 160
Romans, 41–42, 80, 106, 112, 114, 170n50, 174, 197
Rosenberg, Daniel, 11, 78, 127–48, 220
Rotz, Jean, 136, 137
Rousseau, Jean-Jacques, 149, 151, 153, 155, 157, 160, 164
Royal Astronomy Bureau, 71
Royal Society, 78, 127, 133
Rubens, Peter Paul, 108, 115–16
Rumsey, Abby Smith, ix–xi, 220
Rumsey, David, xi
Russia, 23, 56, 58, 60

sacrifice, 81, 91, 101n12, 106
Sadahide, 49–51, 58, 60
Sai River, 54
Saturn, 106
Saxl, Fritz, 108
Say, Thomas, 151
scale, ix, 5, 19, 29, 51, 72, 89, 115–16, 143, 198, 207, 217n38
Schulten, Susan, 11, 147–48, 171–92, 220
Scientific Revolution, 2, 4, 11, 77, 109
Scribner's Monthly, 183
Sedgwick, Adam, 170n50
Sekisui, 43–45, 47
Seneca, 123n36
Seven Years' War, 164
Sherman, William Tecumseh, 181
Shiloh, 202
Shinano, 48, 54–55
shoguns, 35, 41, 47–49, 54, 56–59
"Silurian Beach, The" (Agassiz), 164–65
Silurian System, The (Murchison), 160, 170n50
Silvestris, Bernardus, 106

Simpson, James, 197–98
Smith, William "Strata," 151
Smith, William Thompson Russell, 155
snapshots, 30; animation and, 3–4, 17, 24, 26; archives and, 8; cartographic moment and, 18–25; historical, 15–16, 29; migration and, 183; sequence and, 22, 24, 26, 28–29; temporality and, 3–4, 8, 17–19, 24, 26, 28–29
social constructs, 78–80
Society of Antiquaries of London, 121
Society of Jesus. *See* Jesuit maps
Soderberg, Vicky, 194
Sodom and Gomorrah, 38–39
solar years, 82, 84–85
Song dynasty, 6, 42, 44
Sophocles, 106
Spanish conquest, 77, 80, 93, 97, 100
spatial flow, 17, 25, 27–28
spatial history, 3–4, 14, 16, 25–26, 29, 192n21
spatial memory, 12, 17–30, 29
Spencer, Herbert, 158–59
Stanford Spatial History Lab, 14, 26, 29
static images, x, 3, 8; accommodation of time in, 6–7; Aztec creation stories and, 81; cartographic moment and, 18–25; Jesuit maps and, 73; Rankin on, 9, 16–18, 19–20, 23–25, 27, 29, 32n7, 73; snapshots and, 19 (*see also* snapshots); time maps and, 9, 16–18, 19–20, 23–25, 27, 29, 32n7; Wilkins and, 132
Statistical Atlas of the United States (Walker), 182–84, 185, 188
Stevenson, Seth William, 197–98
St. Louis World's Fair, 203
Strabonis Geographica cum notis Casauboni et aliorum (Wolters), 110–11
strata, 29, 38, 151–57, 159–60
Sugimoto Fumiko, 58–59
Sukjong, King of Korea, 71
synchronism, 103

Tabula Itineraria (Peutinger), 41
Takeda Shingen, 55
taxonomies, 9, 127–30
Teatro della Guerra (Coronelli), 118
telescopic approach, 112, 114, 120
Temple of Time (Willard), 153–55, 176
temporality: analytical approach to, 5; cartographic shutter and, 18, 22, 24, 26, 31; deep time and, 167; defining cartographic moment and, 18–25; digital maps and, 5, 9, 12, 17, 25, 27, 32, 148; dynamism and, 3, 8, 16, 18–19, 25–26, 73, 183; flow and, 3, 17, 19, 21, 24–31; GIS and, 5, 16; historical variable and, 27–31; memory and, 27, 29–30; misconceptions of, 17; new representations of, 12; photo-cinematic idiom and, 17–33; photography and, 18; snapshots and, 3–4, 8, 17–19, 24, 26, 28–29; traditional vs. new approaches to, 17; war and, 215
Tenayuca, 94
Tengu Insurrection, 57
Tenochtitlan, 1, 6–7, 10, 84–86, 89, 92, 93–97, 100
Tenth Census, 184
Tenuch, 93
Tepanecs, 85
Terajima Ryōan, 43
Tertullian, 107
Tezcatlipoca, 81–82, 91
Theatrum orbis terrarum (Ortelius), 39, 41, 67
thematic mapping, 18, 32n10, 179, 184
Thompson, William, 155
Thoreau, Henry David, 104
Tibet, 45, 70
time charts, 161, 175
Time in Space: Representing Time in Maps (conference), ix, 1
timelines, 25, 29, 31, 37, 77, 134, 151, 175, 189
time maps: Bergsonian time and, 26–27; cartographic shutter and, 18, 22, 24, 26, 31; census and, 182; changes in, 8; China and, 45; as cognitive process, 103–4; current events and, 54–60; deep time and, 149–67; defining cartographic moment and, 18–25; durational time and, 3, 5, 10, 13n10, 19, 21; Einstein and, 26, 33n30; grammar for, 134; historical perspective of, 189; historical time and, 1, 3–4, 27, 112, 114, 154; historical variable and, 27–31; Japan and, 45, 54; language and, 134; metaphor and, 103–4 (*see also* metaphor); movement and, 3, 7, 10, 16, 19–26, 31, 65, 81, 84–85, 93, 97, 138, 147, 161, 172–73, 175, 183, 190, 193, 214–15; Newtonian time and, 4, 77; photo-cinematic idiom and, 17–33; postal mapping and, 172, *173*, 182; scale and, ix, 5, 19, 29, 51, 72, 89, 115–16, 143, 198, 207, 217n38; static images and, 9, 16–18, *19–20*, 23–25, 27, 29, 32n7; synoptic gaze and, 103; US origins of, 171–91. See also specific cartographer
Tlaltecuhtli (Earth Lord), 81, 95, 97
Tlatelolco, 89
Tokugawa rule, 35, 41, 47–49, 57–59
Toltecs, 89–91, 94, 97
T-O maps, 38
Tooke, John Horne, *126*, *129*, 132–33, 144n26
topography, 22, 37, 116, 124n52, 157, 172, 187, 196, 198, 200, 207
toponyms, 6–7, 35, 39, 43, 47, 49, 65
Tour from the City of New-York, to Detroit, in the Michigan Territory (Darby), 199–200
tourists: American Civil War and, 196, 202–9, 212–13; automobiles and, 193–94, 196, 206, 209–14; battlefield maps and, 11, 147, 196–215, 216n10; Battle of Waterloo and, 196–98; Grand Tour and, 197, 216n6; guidebooks and, 193–214, 217n20, 217n22, 218n47; Japan and, 49, 58; Lewis and Clark Expedition and, 193–95; Michelin and, 209–10; World War I and, 209–14
Tovar Calendar, 82
travel maps, 193–96, 209–10
Treaty of Guadalupe Hidalgo, 177–78
Treaty of La Mesilla and Guadalupe, 98
Tribute of Yu, 42
trilobites, 150, 164–66
triptychs, 161
Triumph of the Eucharist, The (Rubens), 108, 115–16
Truth, 105–10, 112, 114, 120, 123n28
Truth Being Unveiled by Time (Bernini), 109
Tufte, Edward, 22–23
Turkistan, 70
Turner, Frederick Jackson, 11, 147–48, 184, 186, 187, 189–90
Two general maps of the stars relative to the ecliptic (von Bell), 71

Uesugi Kazuhiro, 51–53
Uesugi Kenshin, 55
Ullman, Edward, 19
Union and Confederate Campaigns in the lower Shenandoah Valley Illustrated (Whitney), 206–7
Union Army, 179–82, 188–89, 203–9, 212, 217n38
United States, 2; *Atlas of the Historical Geography of the United States* and, 22–23, 28, 171, 182, 187–91; census data and, 182–90; coast survey of Civil War and, 178–82, 189; Confederacy and, *180*, *181*, 203, *205*, 206, 209, 212, 217n38; digital technology and, 148, 172, 188–91; Gadsden Purchase and, 98; Great Flood of Mississippi River and, 15; historical atlases and, 11, 172, 174–75, 192n23; Kohl and, 177–79; *Last Leaves of American History* and, 176, 177; Lewis and Clark Expedition

and, 193–95; Library of Congress and, 61n5, 62n24, 64, 74n1, 173, 178, *180*, *183*, *185–86*, 192n12; Lincoln and, 180–81, 202; Mexican-American War and, 98; Mississippi River and, *14*, *15*, 17, 29, 179, 181, 192n13, 194, 196, 199; origins of time maps in, 171–91; population maps and, 182–90, 192n22; postal mapping and, 172, 173, 182; slavery and, 29–30, 174; *Statistical Atlas of the United States* and, 182–84, *185*, 188; Treaty of Guadalupe Hidalgo and, 177–78; Turner and, 11, 147–48, 184, *186*, 187, 189–90; Union and, 179–82, 188–89, 203–9, 212, 217n38; War of 1812 and, 12, 21, 196, 198–202, 212–14; Willard and, 153–55, 173–79, 182, 184, 189, 191n6
Unno Kazutaka, 45
US Coast Survey, 178–82, 189
US Post Office, 172, 173, 182
Utagawa Hiroshige, 56
Utagawa Sadahide, 49–51, 58, 60
Utopia (More), 53

Valley of Mexico, 1, 10, 85, *88*, *89*, 94
van Bleyswijck, François, 111
van Keulen, Jean, 111
Van Leeuwen, Simon, 111
van Leeuwenhoek, Anton, 123n33
van Thulden, Theodoor, 108
Vatican, 118
veils: Baroque style and, 104–5, 108–9, 120; cartographic, 10, 104–5, 115–20; Coronelli and, 105, 116–20, 124n61; discovery and, 104–5, 110–16, 120–21; Dutch and, 105, 108, 111–12, 114, 116, 120, 122n13, 123n27; dynamism and, 105, 110, 120; frontispieces and, 104–5, 109–15, 120–21, 123n33, 123n35; iconography and, 10, 106, 109–11; light/dark dialects and, 105; maps of movement and, 120; metaphor and, 10, 77, 104–6,

110, 118–21; purpose of, 104; of time, 104, 106–14, 121, 122n7
Venice, 10, 105, 116, 118, *119–20*, 124n61
Vicksburg, 179, 181
Vidal de la Blache, Paul, 19, *20*, 25, 27
Vindiciae academiarum (Wilkins), 136
Virtusphere, 138
volumetric mapping, 182
von Bell, Adam Schall, 71

Walker, Francis Amasa, 182–84, *185*, 188–90, 192n22
Walker, James, 217n38
Wang Qi, 43, 68–70, 72, 75n18, 75n24
war: American Civil War, 12, 23, 30 (*see also* American Civil War); Bachelder on, 204–6, 212, 214; battlefield tourism and, 11, 147, 196–215, 216n10; Battle of Waterloo and, 196–98; Chippawa and, 199; Crimean War, 200; diagrams and, 206; guidebooks and, 196–214, 217n20, 217n22, 218n47; iconography and, 210; Lossing fieldbooks and, 200–203, 205, 212, 214; Lundy's Lane and, 199, 214; maps of movement and, 193, 202, 208, 215; Mexican-American, 98; monuments and, 91, 197, 199, 207–8, 210; Napoleon and, 21, 197, 202; Ottoman-Venetian War of the Morea, 118; paintings and, 203, 217n38; Queenston Heights and, 199; Revolutionary War, 164; Seven Years' War, 164; temporality and, 215; tourist maps and, 196, 206, 212; Treaty of Guadalupe Hidalgo and, 177–78; War of 1812 and, 12, 21, 196, 198–202, 212–14; World War I, 12, 24, 196, 209–14; World War II, 29, 77
Ward, Seth, 136
War of 1812, 12, 21, 196, 198–202, 212–14
Warren, Gouverneur Kemble, 179, 192n13
Waterloo, 196–98, 202
Weduwen, Arthur der, 122n13

Western Front, 210
Western Maryland Railroad, 206
Wheeler, George, 179
White, Richard, 16, 25–26
Whitney, Geoffrey, 106
Whitney, William H., 206–7
Wigen, Kären, 1–13, 35, 37–62, 192n13, 220
Wilkins, John: background of, 133; Degérando on, 127–28, 131; diagrams and, 131–43, 145n34; Eco on, 131; *An Essay Towards a Real Character* and, *126*, 127–32, 133, 136, 141; grammar and, 134–35, 140, 144n20, 145n34; language and, 78, *126*, 127–43, 144n13, 144n26, 145n34, 146n45; *Mathematical Magick* and, 138, *139*; Ogden and, 132–36, 144n23; prepositions and, 78, 132–35, 138, 143, 144n26; Ray and, 129; Royal Society and, 127, 133; table of root terms of, 127–28; taxonomy of, 127–30; *Vindiciae academiarum* and, 136; Willughby and, 129
Willard, Emma: "Aboriginal Wandering" and, 175, 189; Bible and, 154; graphic devices of, 175–76; Hart and, 184; *History of the United States* and, 174–75; Kohl and, 177–79; *Last Leaves of American History* and, *176*, 177; mnemonic devices and, 174; Native Americans and, 175; news and, 176–77; "Picture of Nations" and, 175–76; Romans and, 174; *Temple of Time* and, 153–55, 176; time maps of the United States and, 153–55, 173–79, 182, 184, 189, 191n6; World's Education Convention and, 177
William of Orange, 198
Willughby, Francis, 129
Winsor, Justin, 187
Winterer, Caroline, 1–13, 147–70, 220
Withers, George, 103
Wolters, Joannes, 110–11
Wood, Denis, 37

woodblock maps, 41–43, 49, 65, 67–73, 74n1
Woodbridge, William, 174
World War I, 12, 24, 196, 209–14
World War II, 29, 77
Worm, Ole, 112
Wright, Chauncey, 159, 168n7
Wright, John, 16, 22, 28, 33n24, 171–72, 188–91

Xico, 97
Xinjiang, 45
Xuanzang, 38–39

Yellowstone National Park, 187
Yeojido (printed atlases), 72
Yi Gwangjeong, 71
Yokohama, 58, 60
Yosemite, 161

Zenkōji plain, 54
Zhang Hong, 69
Zhang Huang, 67–68
Zhang Wentao, 65
Zhou dynasty, 43
Zhuang Tingfu, 63–66, 69–71
Zolipa map, 97–98
zoopraxiscope, 33n21